Casey Reas ——— Making Pictures with Generative Adversarial Networks

Published by Anteism Books
with Artists + Machine Intelligence

Contents

Foreword

Kenric McDowell and Eva Kozanecka, AMI

Artists + Machine Intelligence (AMI) brings artists and engineers together to realize projects with machine intelligence. Founded in 2016 by Blaise Agüera y Arcas, Distinguished Scientist at Google AI, the program considers art and machine learning together, alongside the emerging relationship between humans and machines.

What does art made with machine intelligence look, sound, and feel like? What can we learn about intelligence, human or otherwise, through artistic collaboration? How can art help us navigate the present and envision our collective future?

As AI becomes part of our daily lives, we must enlist the help of artists and philosophers in designing the technology and our relationships with it. Artists know that a medium or material reshapes the person who uses it, and artists are uniquely able to explore this bidirectional relationship with the material of AI. In doing so, our collaborators bring out broadly applicable insights about machine learning and how we might use it.

In this first non-technical introduction to emerging AI techniques, artist Casey Reas explores what it's like to make pictures with generative adversarial networks (GANs), specifically deep convolutional generative adversarial networks (DCGANs). This text is imagined as a primer for readers interested in creative applications of AI technologies. Ideally, readers will explore the strategies of this emerging field as outlined, and remix them to suit their desires. We hope to inspire future research and collaboration, and encourage a rigorous discussion about art in the age of machine intelligence.

Introduction
Nora N. Khan

My first time writing about Casey Reas was for the *Village Voice* in 2016, after seeing his fourth solo show at bitforms gallery. I was familiar with their storied artistic and pedagogical practice, particularly as the co-creator of Processing, a coding language taught around the world, with its own foundation. I knew how many young artists cited Reas' practice as a touchstone, a starting point for their own growing confidence in making artistic experiments with code. I was further moved to learn that Reas' overall intellectual and artistic approach – open, generous, and inquisitive – matched his dedication to making code accessible.

Back then, when I asked him about the code generating his software paintings, he told me that the code itself was not interesting, but instead, the focus was really the "ideas and systems defined through the code." This imperative, that we try to then articulate those ideas and systems, stayed with me. As a critic, the task of locating the ideas embedded within Reas' *Still Life (RGB-AV A)*, its waves of countless pixels, was daunting. The idea, it seemed, was to first pay close attention to one's own process, as a viewer, of sifting, highlighting, framing, and attributing significance to the output of generative software, executing rules we aren't privy to.

Four years later, I had the pleasure of editing this lucid, concise book by Reas, within the context of Google's Artist and Machine Intelligence (AMI) program. As a handbook, it is exceptional for its accessible walkthrough of the process of training and using generative adversarial networks, or GANs, to create images, that are markedly unlike their source images in form and aesthetic. Reas has described his transitions to GANs as emerging from his practice of creating generative systems and software that make videos and photos. While placing GANs firmly within this lineage of photography, cinema, video, and software, Reas also suggests a new field of aesthetic possibility and evaluative metrics made possible by GANs.

Reas has trained these GANs as sophisticated tools, used in hybrid collaboration with his own intelligence. At the close of this book clearly delineating the technical method of such production, he writes that the resulting pictures should still yet be understood outside of being a product of this process. This is a provocative close. The method and the system are important, but even more critical is the subjective, interpretive space they create for the viewer.

As the artist, Reas firmly takes ownership throughout this picture-making process. He is declaratively uninterested in ceding authorship to the system. In establishing his position on the perpetual question of authorship, Reas helps us follow the entire process as the product of the artist's choice, their agency to select, define parameters, and shape the GAN. The artist remains architect, helping revise and define the GAN. The resulting image's qualities and their impact are made through this system of aesthetic choices.

Reas describes his position within many existing, divergent methods of training GANs, which have a wide range of evaluative measures—from precise fidelity to intentional distortion. He and his studio carefully cultivate diversity in the training photos, working up to a carefully created set, in some cases, of over 200,000 training images. These images might be drawn from a film, such as Ingmar Bergman's *Persona*. Frames of the film fed to train are adjusted, cropped, edited; objects and elements are blurred or foregrounded. Reas insists on a mix of spontaneous choice, taste, and craft needed to ensure there is desired diversity in the training pictures, which has an impact on the generator network's coherence.

Even as this elegant text is unfailingly clear, Reas introduces several gray areas of inquiry from which give us pause, and are jump offs for further investigation.

Whether in photography or working with GANs, the aim, for Reas, is "to make compelling pictures." We start to understand to what pictures made with GANs look and feel like, that the images generated are related to the training images, but "unlike photographs and paintings," Reas writes, are "truly something new." The question of how, is where evaluative inquiry first pauses. How, in fact, are they related and how are they new – in what ways? And how does this quality of newness help us see the original set of images differently? The standard terms of art and film criticism seem to falter here, for how these images might be interpreted.

An answer might be found in closely examining Reas' ongoing series of video works and film stills, made from the outputs of his GANs. The results feel more like abstract paintings in motion. As Reas notes for the GAN trained on *Persona*, a highly cultivated diversity of interior shots, landscapes and portraits give forth images from the latent space that are landscape-skin "hybrids" and some, like the face of the main actress. The GANs' output of images, edited together suggest more of an atmosphere. They form their own cinematic world.

Just as the artifice of film, with its structure and methods, create deeply associative experiences, so too, do the GANs trained on them. The original film's symbolic references begin to emerge. A ghostly figure moves up a hill. A woman melts into a stream. A dais surfaces out of shadows. The ghostly figure continues to ascend hills, confronting some unknown. Candles are lit in a rich, brocaded interior. Patterns in textures coalesce and collapse. Curtains are drawn. A hand hovers above an altar. We sense the seasons passing. The figures melt into their backdrops and environments, melt into each other and cohere. Two people meet in solitude to exchange a message or feeling. Geological change: water levels rise, the land collapses. Fires are lit. We see the ghostly figure on the summit.

The GANs' results are moving. They call on a shared cultural knowledge – whether of film history or myth. The scenes transition here according to a deeper symbolic logic. We get to access a film (that a Reas-made GAN was trained on) through a different lens, an underlying mood that pervades the whole, beyond the narrative, plot, the characters or set design. The visual tensions making up the symbolic world of the film emerge: of man before nature and God, of duality and individualism, or of stable identity against a mirrored shadow self. The eye roams over a loose system of symbolic images that visually represent the dramatic, mythological, and spiritual layers of the whole.

This brings us to the second gray area, that of the viewer's agency in interpretation. Federico Fellini infamously wrote that he didn't believe "rational understanding is an essential element in the reception of any work of art. Either a film has something to say to you, or it hasn't." In his estimation, being moved trumped understanding any process. In the case of using GANs to make pictures, *Making Pictures* strongly suggests that the rational explanation is also deeply critical. In fact, our rational understanding of the process actually helps us enter how provocative this method of image creation is. And with this understanding, we can linger, analyzing, close-reading the image flow for symbols and meaning.

Another gray area is what Reas terms the "ambiguous psychological space," created by the difference between the produced images and its reference images. Take *Persona*, which is in part about the collapse of personalities, visually represented by the collapse of the two main character's faces. Reas notes how the GANs' produced images allow us further insight into this key scene. We have a new sense of how collapse into another person might be visually represented, and felt. What are the new unexpected textures and compositions that emerge, that tell us more about this collapse? Reas invites us to try to imagine the latent space of the original set, with a methodical patience. He asks us to see the whole process and in doing so, take ownership. These are images, after all, which we have rich language for assessing, describing, decoding. We are cognitively primed to navigate any ambiguous psychological space, deploying the "tools" and frames of narrative, historical context, and cultural cues.

A third gray area is the overriding cultural focus on the "dream-like" quality of GANs. There is a tendency to describe their output as subconscious expressions of a network. Reas suggests we can assess this dreamlike space with more rigor. The hybrid forms have a dreamlike quality, but their distortion still depicts the world of the film, and are within legible reading. I interpret the images as a mood or an atmosphere that I associate with *Persona*. There could be an essence of a Bergman or Hitchcock film that is conveyed in these hybrid images, a Bergman-esque, Hitchcock-esque or Brakhage-esque quality that can be surfaced from their films' images, their latent space.

Importantly, Reas' walkthrough eschews any magical language or obfuscation. *Making Pictures* allows us to consider how one might use GANs to create, to teach. The algorithm comes across as having a handmade quality, a tool that is not static, but is refined, reevaluated. Within the generator/discriminator relationship, one side refines and spots forgery and a process of image refinement is undertaken over time. In the context of artistic practice, this process, slowed down, helps us understand our own agency in visualizing and interpreting. Here, seeing and replicating with perfect fidelity is less interesting than coaxing unexpected images from the latent space. The slow transitions allow us to see the field of all possible images, sculpted down through a progression of chosen representations, frames, angles, and compositions. The experience becomes a microcosm of larger scale methods of shaping the world through systems, by naturalizing patterns, hierarchies, and classifications.

Debates in the AI field are often couched in highly technical terms that can be illegible and difficult to access. This language can distance many from feeling they have a stake in discussing AI and deep learning, although their capacities to predict, model, and recreate the world have wide scale social and cultural impact. *Making Pictures* describes AI in a language everyone can understand and engage through. In fact, without such a careful, slow-paced delineation of the process, the AI space quickly becomes one where ideology and power converge in the crafting of seductive illusions. What's often termed the *dreamy, uncanny*, or *surreal* quality of GAN imagery and machine learning's outputs, generally, becomes itself representative of some obscure magic, rather than what we know of it: a process of identifying patterns, pulling out unexpected ones, and training, pushing, and pulling a model with careful curation of training inputs.

Moving through the new images with this concrete understanding, we can read them with a vastly more critical apparatus. Many young programmer-artists today use ML systems to create sets of wild, provocative imagery. But without their verbalized interpretation and framing of the significance of this act, viewers may struggle to discern the significance, as well. Questions remain for future artists training their own GANs: is the process itself—the release of a precious humanist dataset to train a model—the most significant aspect? Is it the resultant psychological space in which viewers try to interpret the meaning of the images? As they embrace their agency in relation to these powerful tools, what new images and patterns can be surfaced? The process, as Reas invites us into it, is replete with chance and discovery.

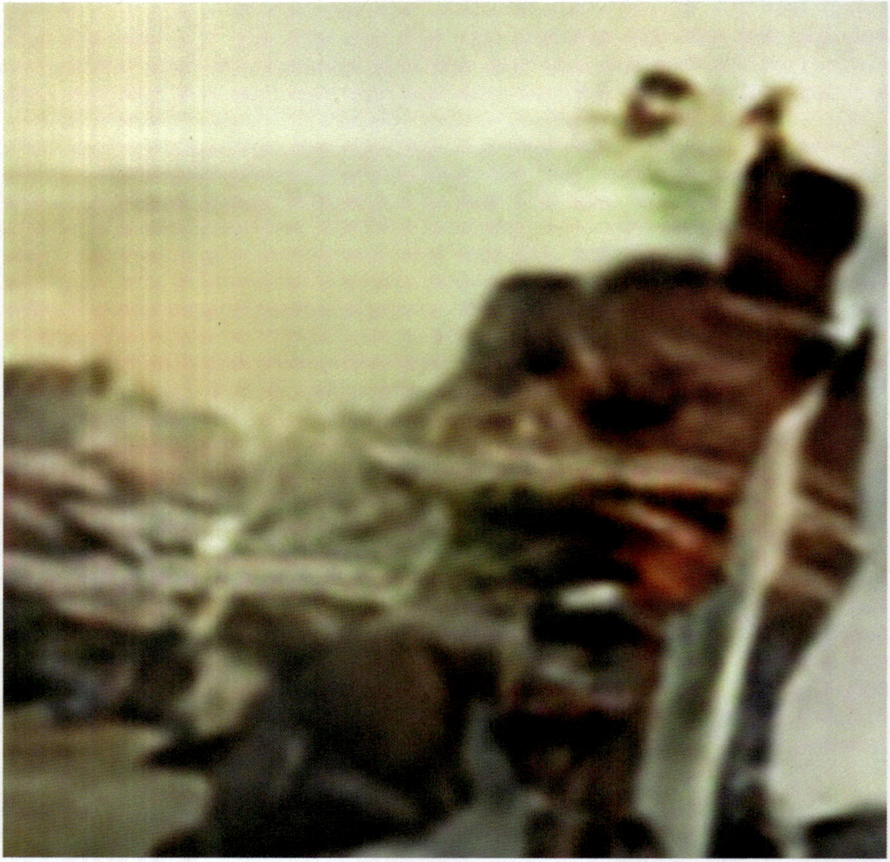

[FIGURE 1]

Picture created with a custom-trained generative adversarial network (GAN).

Making Pictures with Generative Adversarial Networks

In the twenty-first century, media that are thousands of years old are used alongside new ways of making pictures that have only been possible for a few years. At this moment, we might be at the start of a new paradigm in making pictures with machine-learning software that will redefine the visual arts to the extent that photography did so many decades ago. Machine-learning software is different from photography, but there's enough in common to use it as an analogy. In the way that photography was built through and within the sciences of optics and chemistry, so these new media are rooted in computer science, within the domains of machine intelligence and machine learning.[1] Artists don't need to understand the sciences of lenses and chemistry in depth to make photographs, and

[1] The development of lenses for scientific instruments like telescopes led to camera obscuras, boxes or rooms to see the world reflected onto a surface through a lens. Photography used these inventions along with light sensitive materials to fix the images to metal and then later to film.

artists don't need advanced degrees in computer science to work with machine intelligence and machine learning. Machine intelligence and machine learning are wide areas that allow artists to work with language, sound, and pictures in new ways, but here, we are focused on making pictures with *generative adversarial networks* (GANs), specifically *deep convolutional generative adversarial networks* (DCGANs.) A similar essay could be written about many other pathways into the intersections of machine intelligence and the arts.

I am motivated to offer a sense of what it's like to work with GANs to create pictures. If I were writing about photography, I would discuss what it is like to make pictures with a camera rather than to explain how a camera technically operates. If you are interested in GANs from a technical point of view, there are plentiful resources online, and some are referenced in the *Resources* section on page one hundred nine.

There are two discrete steps to working with GANs. The first is to train a model, and the second is to generate images with the model. The model is trained by providing a large set of reference images, typically tens of thousands in number. After the model is trained, it can generate images that are related to the reference images. For example, if a model is trained on one hundred thousand photos of faces, it will generate faces or face-like pictures. It doesn't generate pictures that are the same as any of the reference images. Instead, the model synthesizes the reference images into new pictures. This is what makes working with GANs unique. Training a GAN on faces is only one example. A GAN can be trained on any kind of picture. We'll dig into this by explaining the generative step first, and then we'll move into the training.

Generative

The generative aspect of a GAN utilizes a model to create pictures. To make the comparison with photography, we can also call the model an *apparatus*. Within photography, the camera is the apparatus. The camera apparatus controls the amount of light that comes through the lens, and some cameras can control the focus of the lens. Within analog-film photography, the light sensitivity of the film is a part of the apparatus; with digital photography, this sensitivity is that of the imaging sensor. To get more technical, setting the aperture to f/2.8 and the shutter speed to 1/500 with an ISO setting of 200 will capture a different image than if the aperture is changed to f/16.[2] Changing the physical mechanisms or the software variables of the camera controls the image that is captured.

A GAN model generates pictures by inputting a list of one hundred numbers between -1 and 1. For instance, if all one hundred numbers are set to 0, a specific picture will be produced that correlates to those values. If the first number is changed to 0.1, a similar but different picture will be generated. If all of the numbers are set randomly to values between -1 and 1, a completely different picture will be generated. New pictures are generated by changing any of these numbers. Two similar but unique pictures will be generated with sets of values that are similar. The more different the values, the more different the generated pictures.

[2] The aperture is the hole that light comes through via the lens to reach the light sensitive film or sensor. The shutter speed is how long the light is let in through the aperture before the shutter is closed.

For instance, these are the one hundred values that generate the picture that is in the upper-left corner of **[FIGURE 2]**:

0.7595419088995641	0.8815105587653225	-0.6959948627786567
0.0698149293484449	-0.1724367927190362 2	-0.1081521107779737 1
-0.2523665904248255 7	0.2276612215819888 7	-0.7824200094880887
0.2898852900011608	0.1165049947164089 6	0.7917921176461602
-0.1180221797531133 5	0.0704822105015388	0.9833717688537922
-0.8309206023034232	-0.6857862878613303	-0.2320896416331284
0.3929914803284122 5	0.8383791619492584	-0.3989119297924348 4
0.1079886543208081 8	-0.7974659088486189	-0.5616171353403532
0.7499544496045496	0.8600198447143614	-0.1301356802055655 4
0.978141715252061 7	0.9239518529914019	-0.7257556595856041
-0.945177394849682	-0.836826263808679	0.4925619442015772
-0.0938079125382183	0.9229692925540844	-0.9487774667629587
-0.2263597083180564	-0.3833420544048977 7	-0.0892721872028423 6
0.1179380132200744 7	0.2835836438049786	-0.5718526288111874
-0.7331169069702821	-0.1454936161192837	0.273484709771566
-0.8382659874777176	0.3942342690353209	0.6493137118517 22
0.9289994528676888	-0.196989702087 22	-0.9366866542616381
-0.9147297940811792	-0.8050318064050086	-0.3555515413370049 5
-0.2366186875030305	0.0031960678307700885	0.6813341917993923
-0.1538935779403425 3	0.987167423518491 3	0.0346534841395584 9
0.0957594731999329 4	-0.1695220444868657	0.5169669751867703
-0.817237409433499 1	-0.3825240248627068 7	0.2934740055866944 5
-0.1120412326975497 3	-0.9685925109515146	0.6193855535994541
0.5850419114263421	0.9663115997451939	-0.0657406446261030 7
-0.991027925107417 2	-0.686215781922958	-0.5706316835475198
-0.1589578955615849	-0.3840338904001805 4	0.1888183458623087 2
0.5530314419251254	0.0326602694014668 24	-0.2446131808148994 7
-0.7688423816791075	-0.3862246037174965 6	0.3823891305933631
0.8306847102146822	-0.9134488435163859	-0.3197687445936268
0.811322776604448	-0.2380696643649524 7	-0.5245023830075208
-0.955797954908187 9	-0.8579780062416458	-0.3648358039978738 5
0.9105277485301015	-0.7308220808647958	-0.2772681943032633
-0.6278208303304886	-0.4148126337968032	
0.720209173407359	-0.0824215153004717	

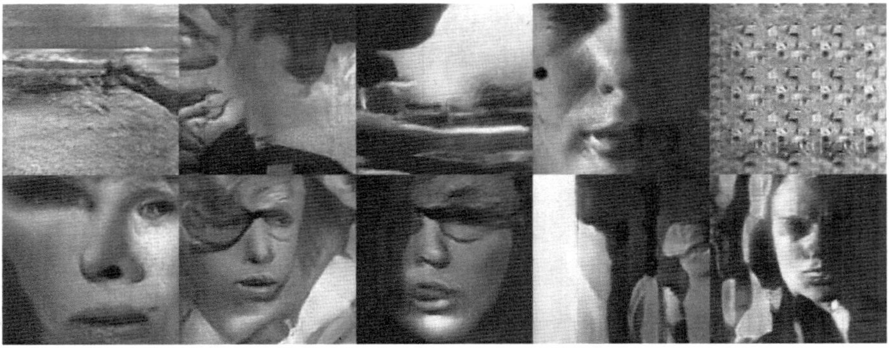

[FIGURE 2] **Sample images generated from a DCGAN model trained on a black and white film.**

These values can be visualized in different ways to see the numbers as images. For example, it's easier to get a sense of these values by looking at them as grayscale bars where the value -1 is black and 1 is white with continuous gray values in between. When these numbers are converted using a short Processing program, this is the result:

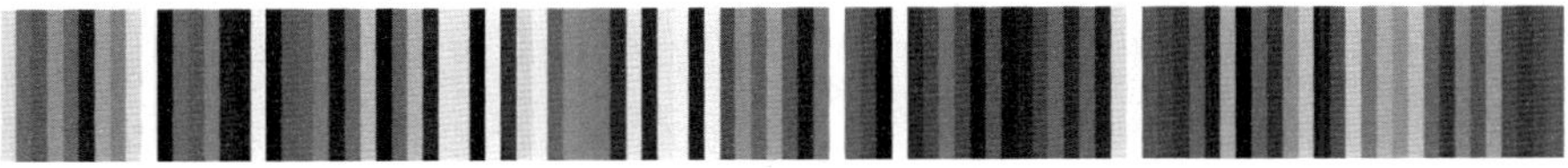

To create new pictures from a DCGAN model, change the numbers and generate again. The grid of pictures in [FIGURE 2] are generated by training a DCGAN model on frames of *Persona*, a film directed by Ingmar Bergman, and released in 1966. Some of the pictures are recognizable as the actors from the film. Others are distorted beyond recognition, and others are more abstracted. Some pictures of faces don't feel like any of the actors in the film; they are hybrids and new forms. Other generated pictures are uncomfortable combinations of landscapes and skin texture. These images have an uncanny and dreamlike quality. In general, images created by GANs often have a surreal tone. In the case of the film *Persona*, these distorted images depict the visual world of the film differently than the original images captured through the camera lens. As the film progresses, the distinct personalities of the two characters dissolve into an ambiguous psychological space that culminates with the faces of each character superimposed into a single visage. A subset of images created through the GAN are an alternate way to imagine this essential aspect of the film.

A different DCGAN model, trained on another set of sample pictures, produces other types of pictures, but the apparatus is the same. **[FIGURE 3]** is a sample of pictures generated from a model trained on color scans of plants gathered in Colorado as part of the *Study for a Garden of Earthly Delights* series. From a set of fifty high resolution scans, one hundred thousand pictures were cropped and used to train the GAN.

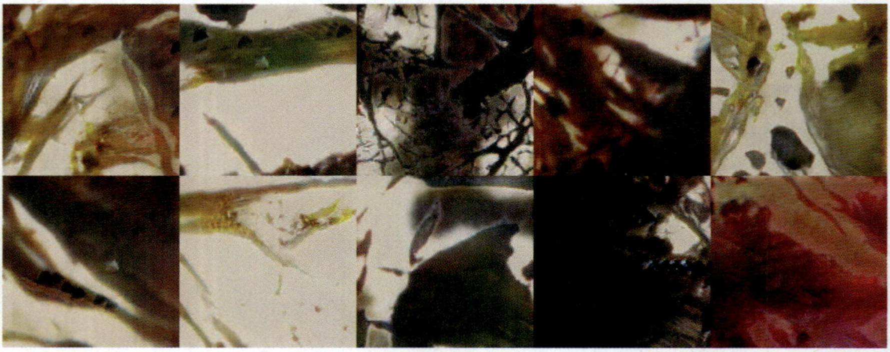

[FIGURE 3] **Sample images generated from a DCGAN model trained on color scans of plants.**

To directly show the similarity, the numeric values below generated the image in the upper-left corner of **[FIGURE 3]:**

-0.7403904593033481	0.1034479856432226 6	0.8036241216688109
0.9432629637114365	-0.1708367544356017 6	-0.4462814263923917 7
0.4474684723996884 5	0.8419651643503598	0.8313690829568507
0.554208968753104	0.2766866346703489 3	0.6265973579848199
0.0718419471655305	-0.0814110185491616 6	-0.0250999718246731 62
-0.8005537720672147	-0.245874132575462 3	0.1727865090824194 4
-0.8532894021665063	-0.2742846880800642 3	0.7576426037752679
-0.816930160235841 6	-0.2489804744956605 2	-0.7158761575366988
0.0914090264643969 7	0.3141143943949873 4	0.6112987379236214
0.164315924441353	-0.195863459268933	-0.5857655415436003
0.6307445989149931	0.4158161020633075 5	-0.5286279955952804
0.6079271580512469	-0.286333599427796	-0.6196269173409219
0.6210589775877251	0.2235224902606978 1	0.1772283154666605
-0.4421683097389264	-0.8139242561087905	-0.5517284211752875
-0.3509061040352133	-0.4369978676719463 7	0.2973479063848989 4
0.0111737465840606 73	0.6861436206583973	0.3417956885373241
-0.7568597529879437	-0.3810780998186396	0.2894629742665386 6
-0.9341244228786985	0.5211485426841798	0.1068813462539330 8
-0.8556153694141084	-0.6698500704796888	-0.277173219232848 9
0.2903263337763635	0.9101806577024842	-0.1139008088608817 2
0.4247736996210731 4	-0.1618410992644885 3	-0.3740195475833296 7
-0.3368435379562089	0.9750045774767491	-0.271571371744435 7
-0.4391242407253031	0.7756690864853439	-0.0706922651293864 2
0.7592859050988334	-0.5661429228546253	0.816164611685 19
-0.8797235112062911	-0.6929441772883493	0.6114554646898565
0.8540958258355191	0.1286490882613056	0.8700914868476461
0.1450135136495134 3	-0.6839434914567546	-0.2174478761224689 6
-0.9076784027924942	-0.3758219031025798	-0.2640143624736033
-0.3933857075339557 4	0.4865347634798487	0.4213137267723144
0.4830898366514542	-0.9947514752073525	0.1056955042598071 2
0.4822857085686471 6	0.6035889121267559	0.7959487455586816
-0.799275404173592 3	0.1411293394698509 6	-0.6141287538444893
0.8747327185682279	-0.268575221249482 3	
-0.2368242467398802 3	-0.4678774346392718	

This set of numbers can also be converted to grayscale values as an alternate representation:

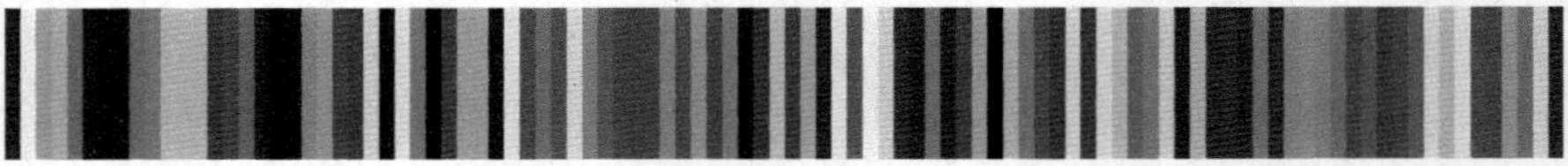

The knowledge that a set of numerical values between -1 and 1 map a specific picture leads to new questions. Why do the numbers correlate to specific pictures? How do precise changes to the numbers manifest as changes to the pictures? Are all of the possible pictures similar to the sample pictures, or do unexpected textures and compositions emerge?

Latent Space

Until pictures are generated with a DCGAN model, there's no way to know what will be created. The list of one hundred numbers seen above is called the *latent vector*, where the word "latent" means *hidden*. Each set of possible values is a map to generate a unique image within the entire latent space, the space of all pictures that can be generated from the model. We can move—interpolate—through the latent space to reveal the pictures. The latent space can be navigated at random to sample the entire space as shown in **[FIGURE 4]**, but it can also be navigated with order to generate a series of related pictures. If the values are close together, the transitions are smooth, but when the values are further apart, the transitions are more abrupt, as seen by comparing **[FIGURE 5]** and **[FIGURE 6]**.

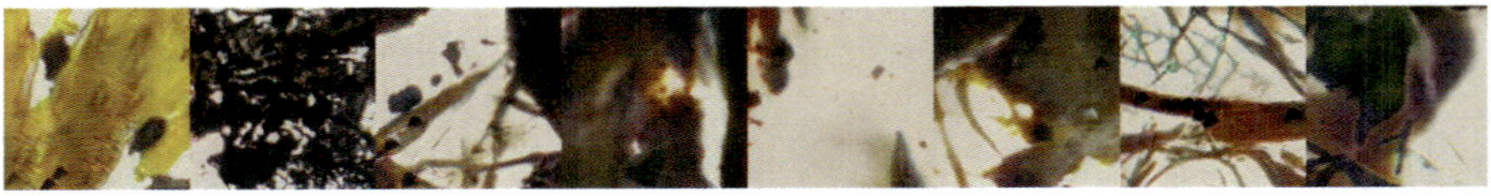

[FIGURE 4] **Selecting positions from the latent space with random numbers generates unrelated pictures from throughout the latent space.**

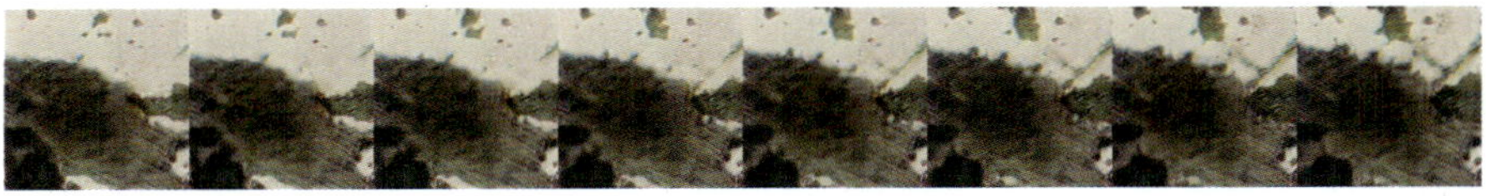

[FIGURE 5] **Moving through the latent space with small changes made to the values creates a slow transition.**

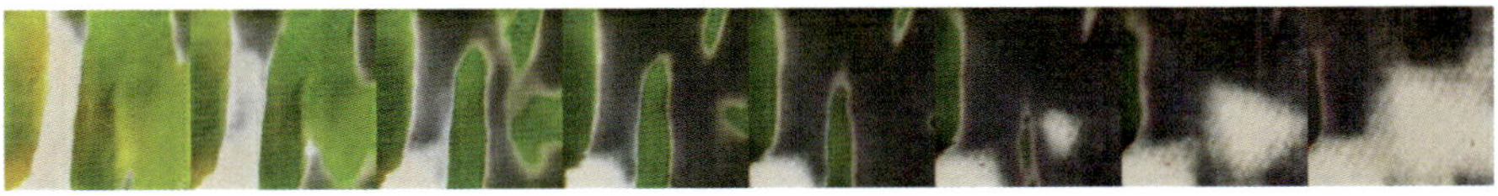

[FIGURE 6] **Moving through the latent space with larger changes made to the values than [Figure 5] creates a faster transition.**

When the images are sequenced in time, as in videos or animated GIFs, moving through the latent space creates animation. Like with [FIGURE 5] and [FIGURE 6], the space can be navigated either smoothly and slowly, or quickly and erratically. It is also possible to move through the space in both ways at the same time, and to choreograph these changes to create cuts and slow transitions. Once the DCGAN model is trained, the way it is utilized for animation is a result of choreographic and visual thinking. [FIGURE 7] and [FIGURE 8] show two options for moving through latent space and sequencing the pictures to create animation. These videos frames are raw; they are created directly from images generated by the trained model.

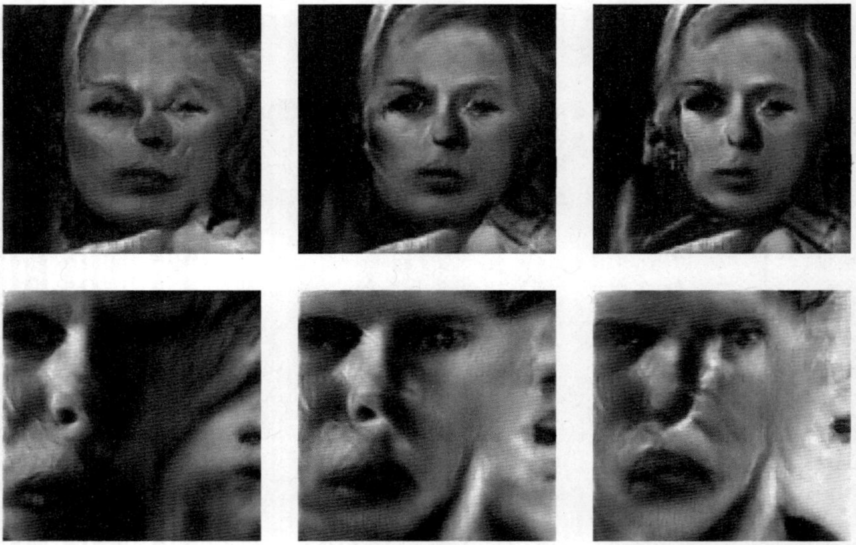

[FIGURE 7] **Animation created by exploring the latent space of the Persona DCGAN training. See [Figure 2] for sample images from this model.**

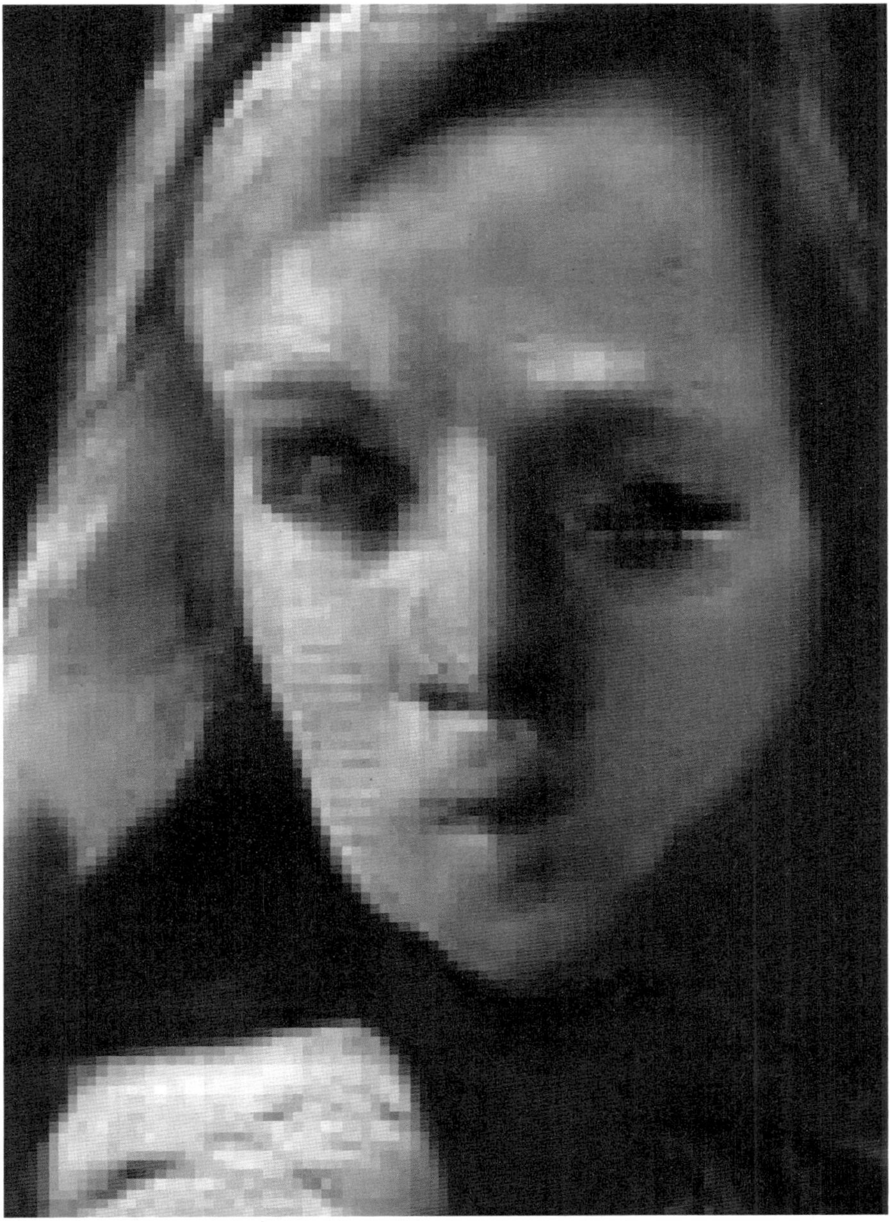

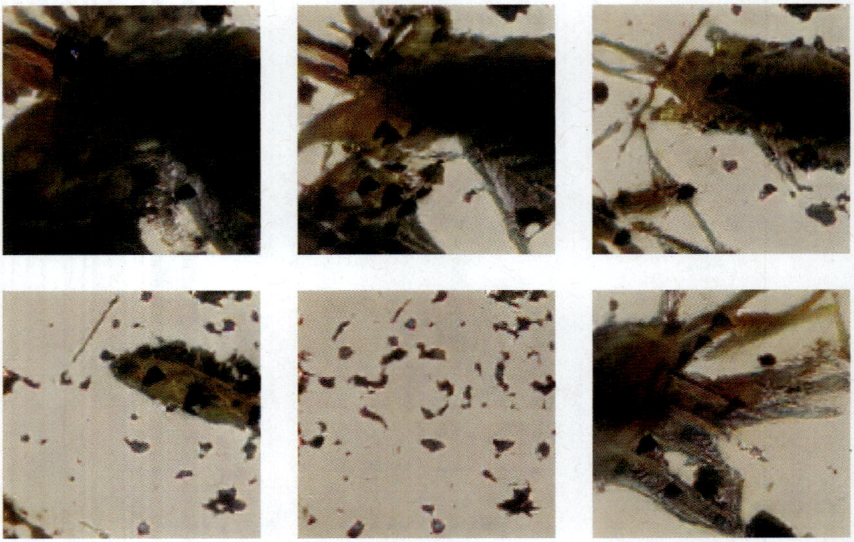

[FIGURE 8] **Animation created by exploring the latent space of the *Study for a Garden of Earthly Delights* DCGAN training. See [Figure 3] for sample images from this model.**

The animation in **[FIGURE 9]** is similar to that in **[FIGURE 7]**, but each frame of the animation was adjusted. They were blurred to remove artifacts of the GAN, enlarged and converted to grayscale. Then, a small amount of digital noise was applied. **[FIGURE 9]** points to the opportunity to continue to work with the images after they are generated. Generating an image with a GAN can be thought of as the start of another process, in the same way that capturing an image with a camera is often only one step in the larger system necessary for making a picture.

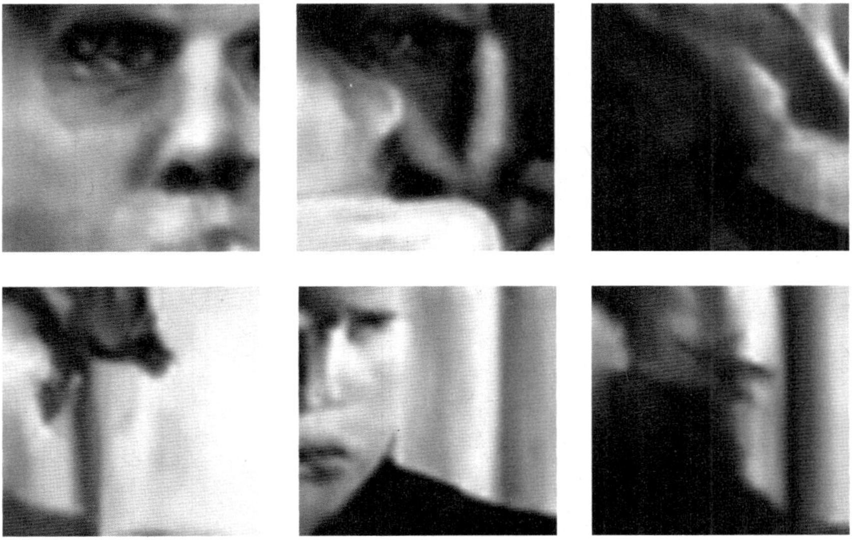

[FIGURE 9] Animation created by interpolating the latent space of the *Persona* DCGAN training. Each frame of the animation was modified by changing the scale, averaging the pixels, and adding noise.

The Ground Truth

In photography, selecting where to take photos and what to take pictures of are major decisions; these may be the most significant decisions a photographer makes. Are you like Gordon Parks shooting at a civil rights rally, or like Barbara Kasten composing in the studio? The equivalent decision in working with GANs is selecting which pictures to train with. Consider two contemporary examples of working with different kinds of GANs: artist and human computer interaction (HCI) researcher Harshit Agrawal, who trained on sixty thousand pictures of surgical dissections for *The Anatomy Lesson of Dr. Algorithm;* and artist Anna Ridler, who rendered two hundred ink drawings to train for *Fall of the House of Usher.* Agrawal chose to place the work in the tradition of paintings like Rembrandt's *The Anatomy Lesson of Dr. Tulp,* created in 1632, that renders the interior of the body with uncompromised detail. Ridler's work is focused in a different direction, toward her own ability to draw and to explore how the GAN analyzes her specific marks in a different way to manipulate a medium that is "cold, sterile and algorithmic to maintain and accentuate a sense of human touch."[3]

Creating these large sets of training images is one of the most time consuming and challenging aspects of working with GANs. My experience with DCGAN is that training with 128 × 128 pixel resolution pictures requires tens of thousands of images. The best results have come from training with one hundred thousand images or more. For instance, when training the *Study for A Garden of Earthly Delights* GAN in [FIGURE 3], the process did not work with five thousand images, as can be seen in [FIGURE 10]. The generated pictures remained fuzzy and without clear structure; with one hundred thousand sample images, however, the training worked well. The generated pictures felt like examples from the sample pictures.

[3] As described by the artist on her website: http://annaridler.com/fall-of-the-house-of-usher/

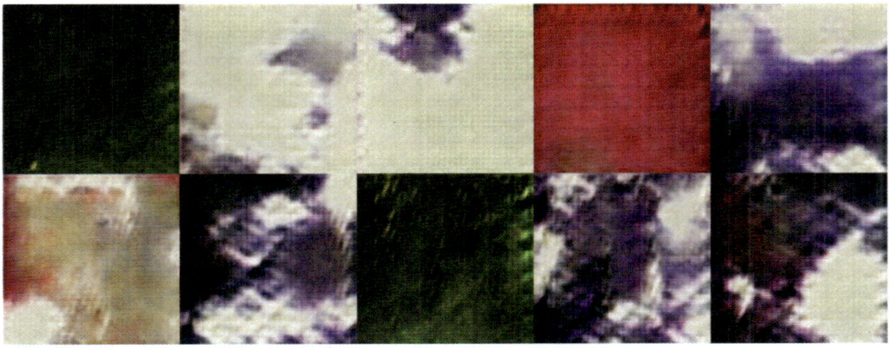

[FIGURE 10] **Examples of generated pictures from an unsuccessful DCGAN training, with only five thousand reference pictures from *Study for a Garden of Earthly Delights*. Compare this to [Figure 3], which uses the same kind of reference pictures for training.**

Once a set of pictures is ready, the training can begin. Depending on the quantity of sample pictures, the parameters of the training, and the computational resources, the training can take between a few hours and a few days or weeks. While the model is training, it is common to generate reference images at fixed intervals to see what is happening with the training.

Neural Network

Before discussing the idea of the *adversarial part* of the *generative adversarial network*, it's important to introduce the idea of a network. Here, the term "network" is condensed language referring to a *neural network* or *neural net*. A neural network is a network of *artificial neurons*. The idea for "neuro-logical" networks dates back to the 1940s.[4] The driving question was, if the way the brain learns can be simulated by a machine, can a machine learn like a brain? These ideas contributed to the formation of *cybernetics* in which organic and electromechanical systems began to be integrated as the study of "control and communication in the animal and the machine."[5] Research into neural networks moved from theory to practical application. Despite fallow years, the research is currently thriving.

The *deep* part of a *deep convolutional generative adversarial network* means the network has more than three layers. The first layer passes information to the next layer, which passes information to the next layer, and so on. To be more precise, with a DCGAN model, the one hundred numerical values (as seen above) are expanded layer by layer. If the model was trained with 128 × 128 pixel images, the one hundred numbers eventually expand to 49,152 numbers, which is calculated as 128 × 128 pixels × 3 color values for the red, green, and blue components of each pixel.

[4] For example, see "A Logical Calculus of Ideas Immanent in Nervous Activity," published in 1943 by Warren McCulloch and Walter Pitts.

[5] Norbert Wiener's influential book *Cybernetics: Or Control and Communication in the Animal and the Machine* crystallized the domain of cybernetics.

For a machine learning researcher, the goal of training a GAN is to compare the fidelity of a decoded image in relation to the original images in the training data, and to see no distinct differences. For visual artists, the goal might be similar, but they might also strive for differences between the training data and the generated images. Along these lines, the crux of working with GANs after they have been trained is to manipulate the one hundred numbers to coax unexpected images from the latent space, images that are not clearly associated with the training pictures, (but which *feel* like they are related to different degrees according to the aims of the artist). A neural network can be used to explore and discover pictures that are different from what the artist can create without them.

Adversaries

The innovation of GANs in relation to other kinds of neural networks is the *adversarial strategy* to train the model. This means there is a pair of networks that are trained in competition with each other. One network is named the *generator* and the other is the *discriminator.* The generator is the part of the model that generates images from the set of one hundred numbers. A common visual arts analogy refers to the generator as the *forger,* and to the discriminator as the *expert* that determines the authenticity of the image produced by the forger.[6] The discriminator/expert has access to the picture created by the generator/forger as well as the sample pictures in the training data. The role of the discriminator/expert is to declare if the picture made by the generator is a "forgery" or not, and it only has two options, to answer "yes" or "no." The goal is for the generator to be so good at creating "forgeries" that the discriminator can no longer reliably distinguish if the generated picture is one of the sample pictures or not. Over time, the discriminator "guides" the generator to learn how to make pictures that the discriminator can't detect as "forgeries."

The generator and discriminator are adversaries, each trying to outperform the other. They are both trained; they both "learn" how to perform their roles. What is the generator learning? What is it being trained to do? It is learning to estimate the probability distribution of the data from the pictures in the training set. The practical outcome of this is generating pictures that feel like the training pictures, but don't exist within the training-pictures set. The generator creates new pictures and the discriminator can't tell if they are "real" or a clever "forgery." The discriminator learns to be better at evaluating the differences between the sample pictures and the pictures created

[6] This analogy was discovered through the paper "Generative Adversarial Networks: An Overview," by Vincent Dumoulin et al.

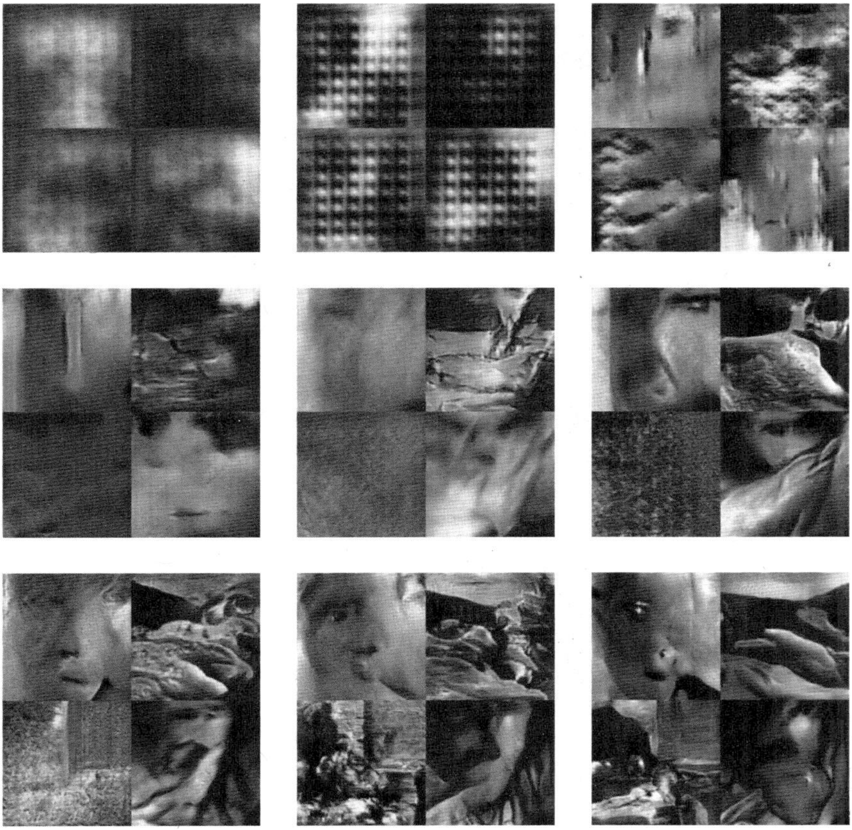

[FIGURE 11] Timelapse showing the development of the *Persona* DCGAN model from epoch 0 to epoch 24. Each square area of the picture is the result of the same set of one hundred numerical values fed into the GAN, but the picture changes as the GAN is refined.

by the generator. The generator has access to the information about why the discriminator thinks a generated picture is fake or not, and it uses that feedback to correct itself.

The diversity of the training pictures has a major influence on the outcome. If the images are extremely diverse, there's not enough of a pattern for the generator to learn and the results are incoherent. Sometimes this is desired and sometimes not. If the images are too similar or too few, the model might not train at all, as seen in [FIGURE 10]. The goal is usually to have a balance between homogeneity and diversity in the training pictures. For example, the pictures used to train the model featured in [FIGURE 3] had diversity of texture, shape, and color, but they all had similar organic form and were flat, because they were all captured with a scanner. The pictures used to train the model in [FIGURE 2] were more diverse. Consider the film that this range of pictures was sourced from. *Persona* has two primary actresses, who are filmed from every angle and distance from the lens. Furthermore, the film includes portraits, landscape, and interior shots. Because of this greater diversity in the training pictures, there are more uncanny hybrids in the generated pictures. Some locations in the latent space look strikingly like Bibi Andersson, one of the actors, and other locations reveal themselves, with closer analysis, as hybrids of skin texture and landscape form. To train using captures from *Persona,* each training picture was a cropped enlargement of each film frame, to increase the diversity of the images. When each frame was used "as is" for the training, there wasn't enough diversity to train the model.

When the training is finished, how is it evaluated? What does a good training mean? I think this qualification is ultimately defined by the artist. Some artists seek and desire fidelity, while others seek a specific kind of

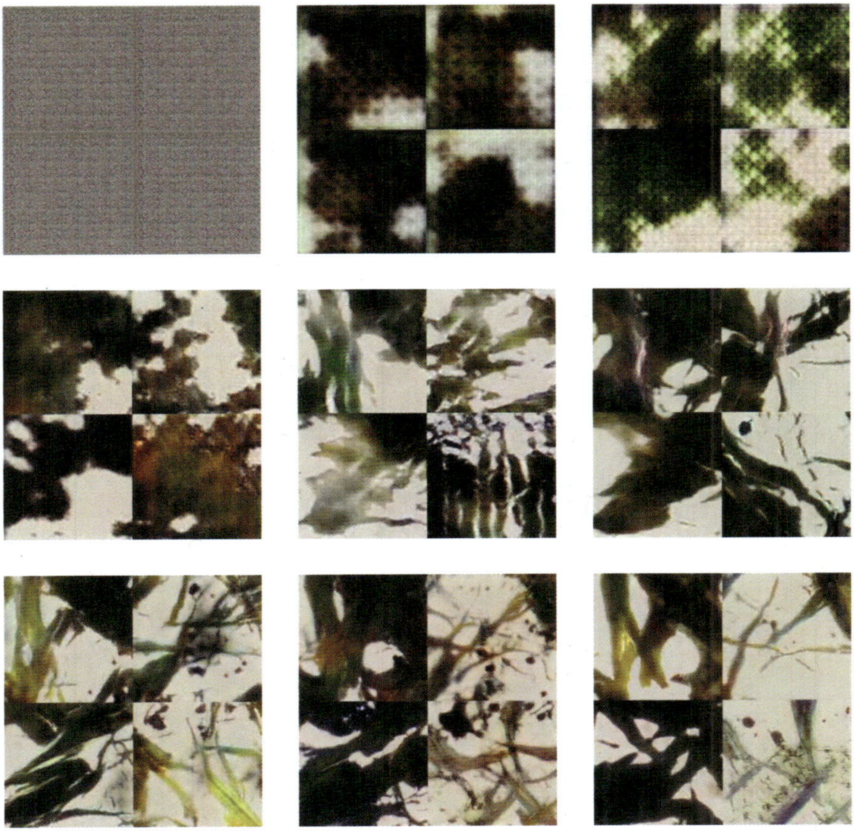

[FIGURE 12] Timelapse showing the development of the *Study for a Garden of Earthly Delights* DCGAN model from epoch 0 to epoch 24. Like with [Figure 11], each square area of the picture is the result of the same set of one hundred numerical values fed into the GAN, but the picture changes as the GAN is refined.

distortion. There are, of course, divergent goals for craft within each form of the visual arts, from painting to sculpture to photography. The same is true in working with GANs. One person's idea of a poorly trained model might be another's desired result. The aim is to make compelling pictures, not to demonstrate technical ability.

Process

The agency of an artist working with GANs is in selecting the training pictures, defining the parameters of the training, and editing the generated images. Much as in photography, there are often hundreds or thousands of pictures to work with at some point in the process. Unlike in photography, there aren't a fixed set of frames with GANs. It's nearly always possible to interpolate between pictures to generate something in between, or to push and pull a picture into different directions because the latent space is continuous.

After pictures are generated from a trained GAN model, what happens next? Is there another step of further transformation and refinement? If I am working with black and white chemical photography, it is time spent in the darkroom enlarging, cropping, dodging, and burning. With digital photography, it is time spent with software modifying the values of pixels. What is the equivalent when working with GANs? So far, in my experience, the process is more like working with digital photography than anything else. I spend time in image editing software changing the color balance, resizing images, and cropping them.

One can also use GAN pictures as reference images, rather than as the end result. Like painters who use photographs as source images, GAN pictures can be the start of work to be completed in another medium. For instance, as early as the end of the nineteenth century, the eminent American painter Thomas Eakins used photographs that he took as references for his most well-known paintings.

Like every medium, GANs have limitations. You might have noticed that the majority of the figures in this essay are square and low resolution; these are the defaults. Pictures generated with a DCGAN model have a shockingly

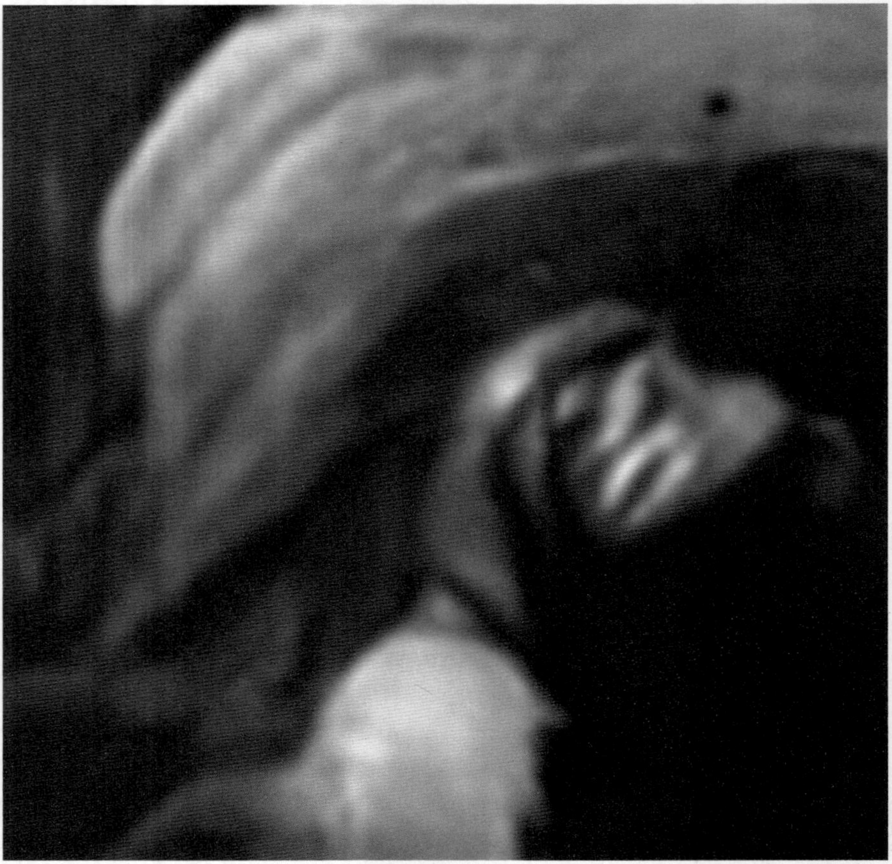

[FIGURE 13] Picture created with a custom-trained generative adversarial network (GAN).

low resolution by current standards. It is easiest to train a GAN to work with images that are 64 × 64 pixels. It gets more difficult as the dimensions increase. So far, in my studio, we've managed training with 128 × 128 pixels and we plan to soon make tests with 256 × 256 pixels. There are other kinds of GANs built for enhancing resolution. Pioneers like Mario Klingemann and Mike Tyka connect the input of one kind of GAN to another, to increase the resolution. The ability for GANs to work with larger images is quickly developing. Within a few weeks of writing this, there will likely be new methods and available code. This is a rapid area of research and there are regular research breakthroughs published as new papers with associated code that make it possible to work at higher resolutions and to create images that feel less artificial. This work is technically challenging, but I am deeply engaged with the new forms of pictures made possible by working with GANs. In the studio, we meet the technical challenges as needed to focus on the pictures.

Pictures

New types of art and new ways of thinking emerge from new technologies. Software as a medium for visual art grew out of video, which was developed from the moving-image space of film and prior to that, from photography. Each new medium defines new paths for exploration that absorb established media and imagine something new. As emerging media develop, the influence becomes bidirectional. As a simplified example, think about how printing presses changed the form and economy of the book, which later led to new forms of writing, distribution, and wider ideas about literacy. Here, the focus is on GANs and at this point, we don't know what kind of impact they will have on the visual arts and culture at large. At this early stage, it's convenient to discuss photography because knowledge of photography is widespread and it's a bridge to the unknown space of GANs. They will likely evolve so this comparison will be less applicable.

There are always questions about authorship when software is involved in the creation of work, and these questions are accelerated when machine intelligence and machine learning are involved. Questions ripped from the headlines of the day such as "Can AI be creative?" and "Is the AI the artist?" are orthogonal to my interests. I have no desire to cede authorship to a system. I think of these emerging software systems as extensions of my own abilities and I maintain complete authorship. My engagement is with the fundamental ideas, and with the ability for the software to assist in creating pictures. The pictures are the complete focus. Ideally, they can be approached and interpreted outside of the orbit of any specific method of production.

Untitled Film Stills 2.1 – 2.24

Each *Untitled Film Still* image is a frame from an imagined film. The full set of images defines a more complete perspective of the possible scenes. The small image size, 30 × 30 cm, invites close proximity for viewing and encourages attention toward the details. The soft focus and grain are reminiscent of early explorations into photography and an adjustment to the low resolution of GAN images in 2019.

These images were created by training a DCGAN model on over 100,000 small images (128 × 128 pixels) that were resized and cropped from every individual frame of a selected film. The final set of 24 images were selected from thousands of randomly-generated latent vectors and their corresponding images.

Untitled Film Stills 2.1 – 2.24 were first exhibited at the DAM GALLERY in Berlin in March 2019.

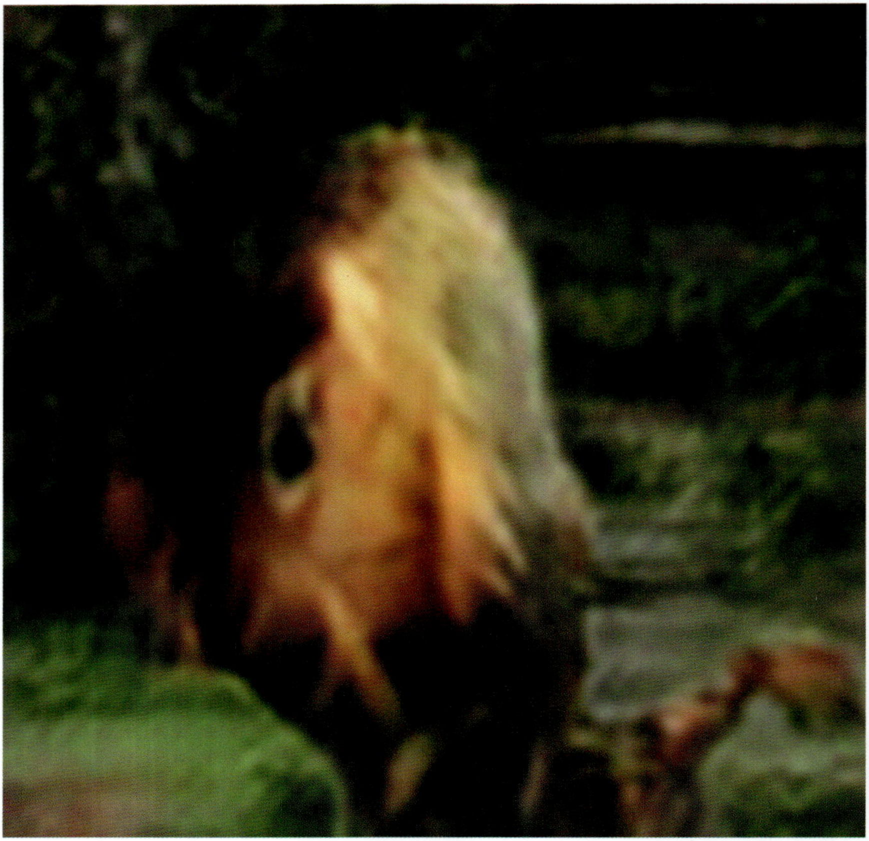

Untitled Film Stills 2.1 2019
C-print
11.8 × 11.8 in / 30 × 30 cm

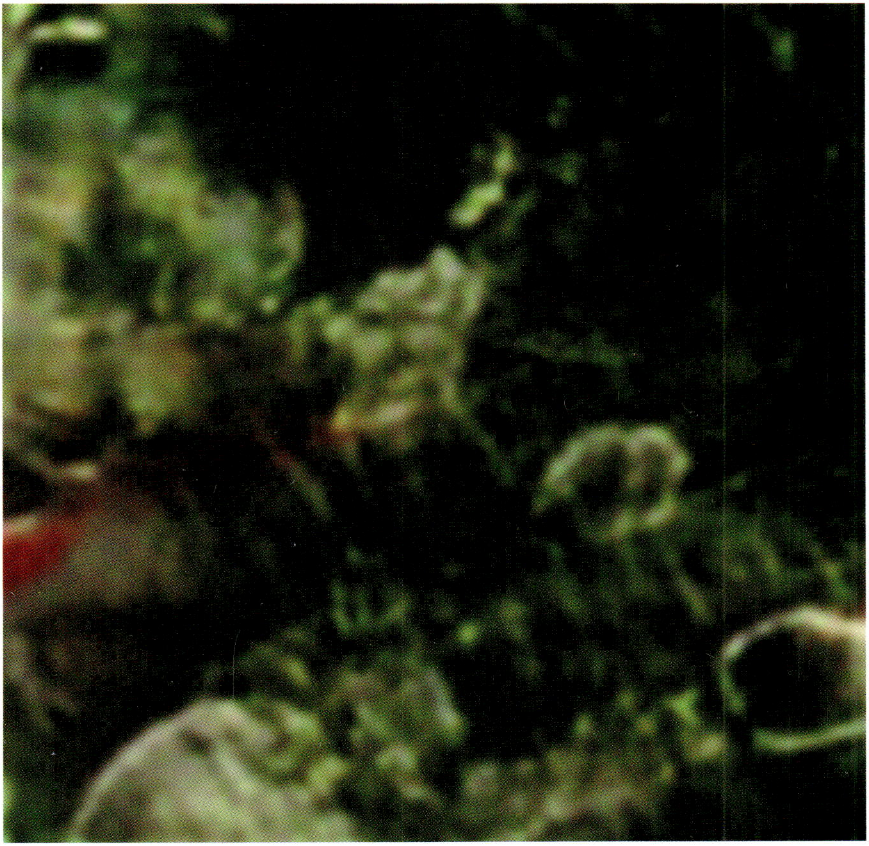

Untitled Film Stills 2.2 2019
C-print
11.8 × 11.8 in / 30 × 30 cm

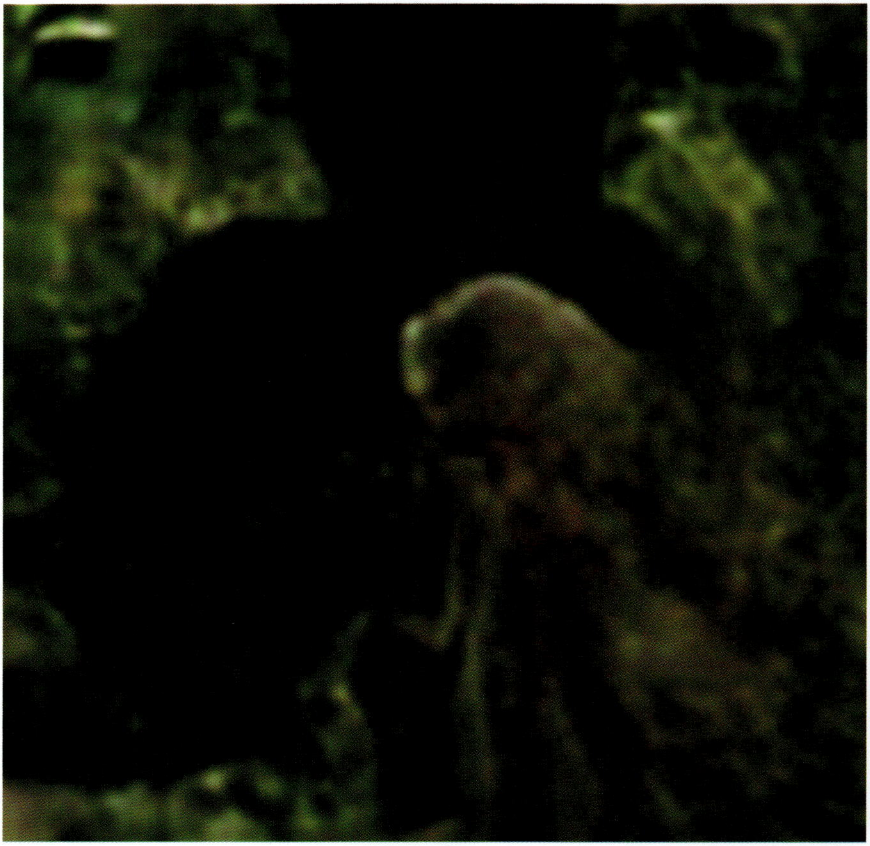

Untitled Film Stills 2.3 2019
C-print
11.8 × 11.8 in / 30 × 30 cm

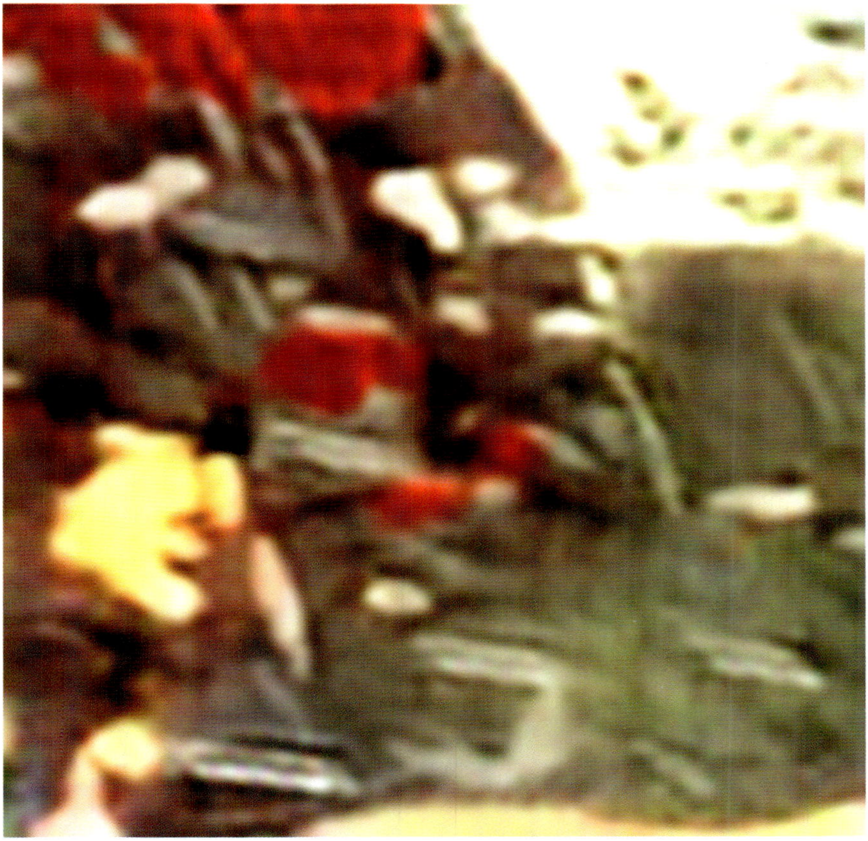

Untitled Film Stills 2.4 2019
C-print
11.8 × 11.8 in / 30 × 30 cm

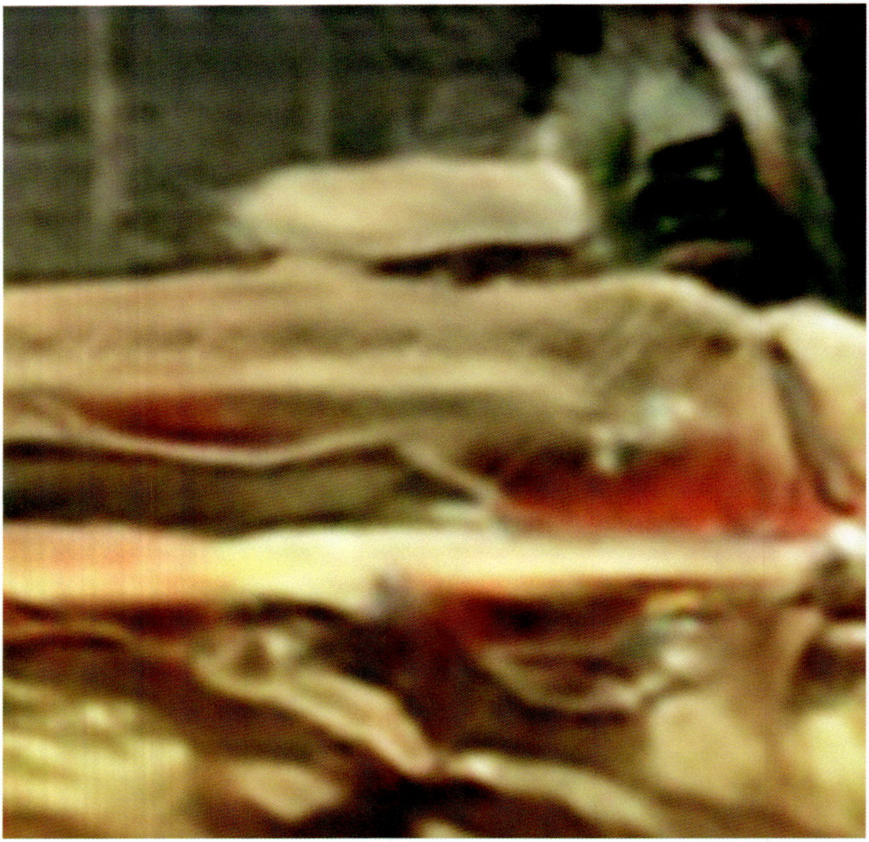

Untitled Film Stills 2.5 2019
C-print
11.8 × 11.8 in / 30 × 30 cm

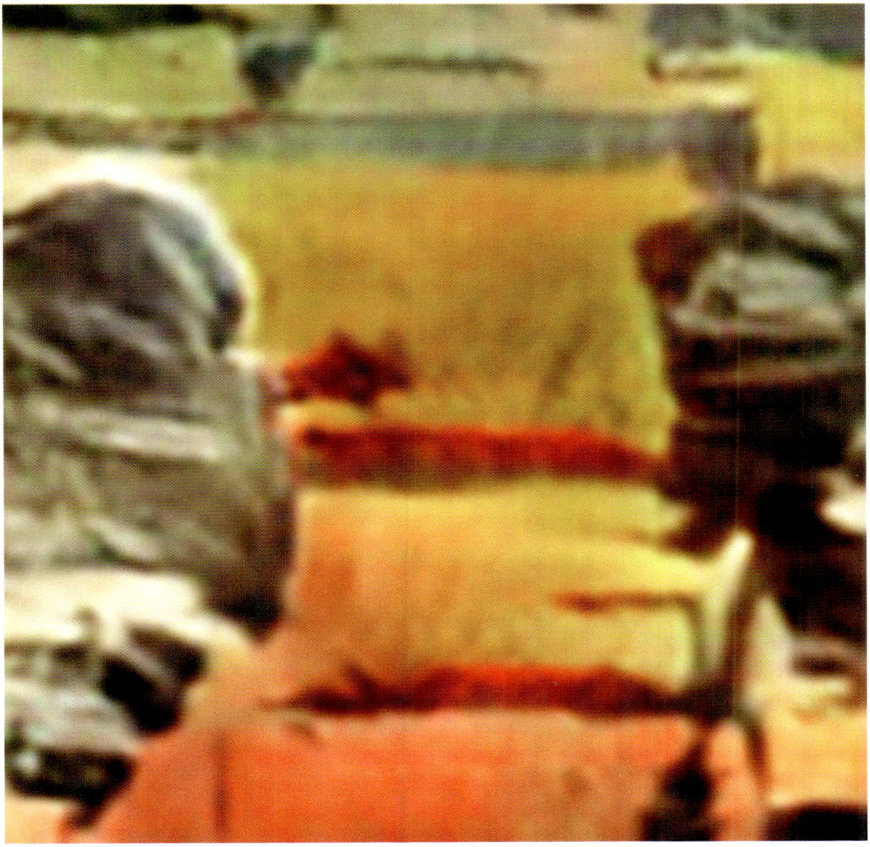

Untitled Film Stills 2.6 2019
C-print
11.8 × 11.8 in / 30 × 30 cm

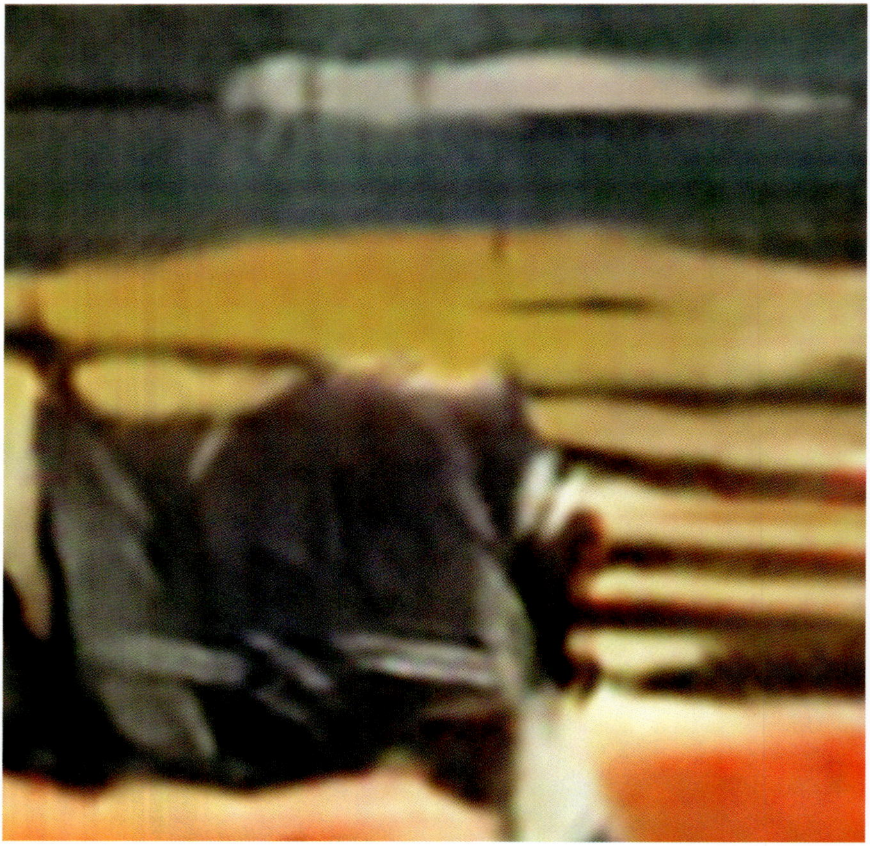

Untitled Film Stills 2.7 2019
C-print
11.8 × 11.8 in / 30 × 30 cm

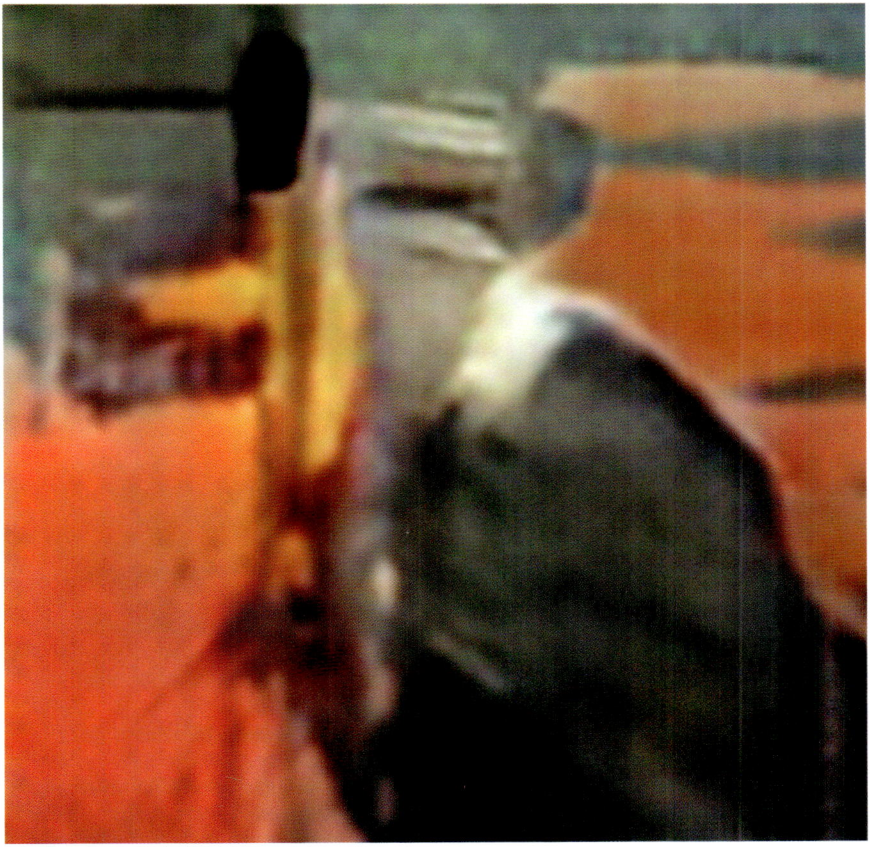

Untitled Film Stills 2.8 2019
C-print
11.8 × 11.8 in / 30 × 30 cm

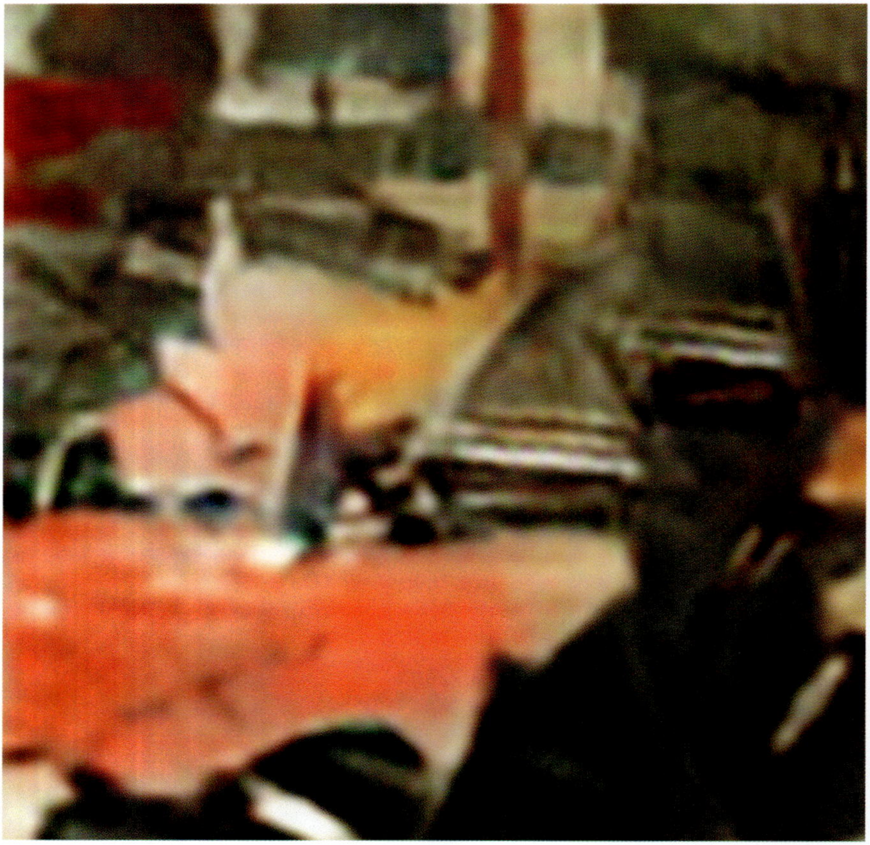

Untitled Film Stills 2.9 2019

C-print

11.8 × 11.8 in / 30 × 30 cm

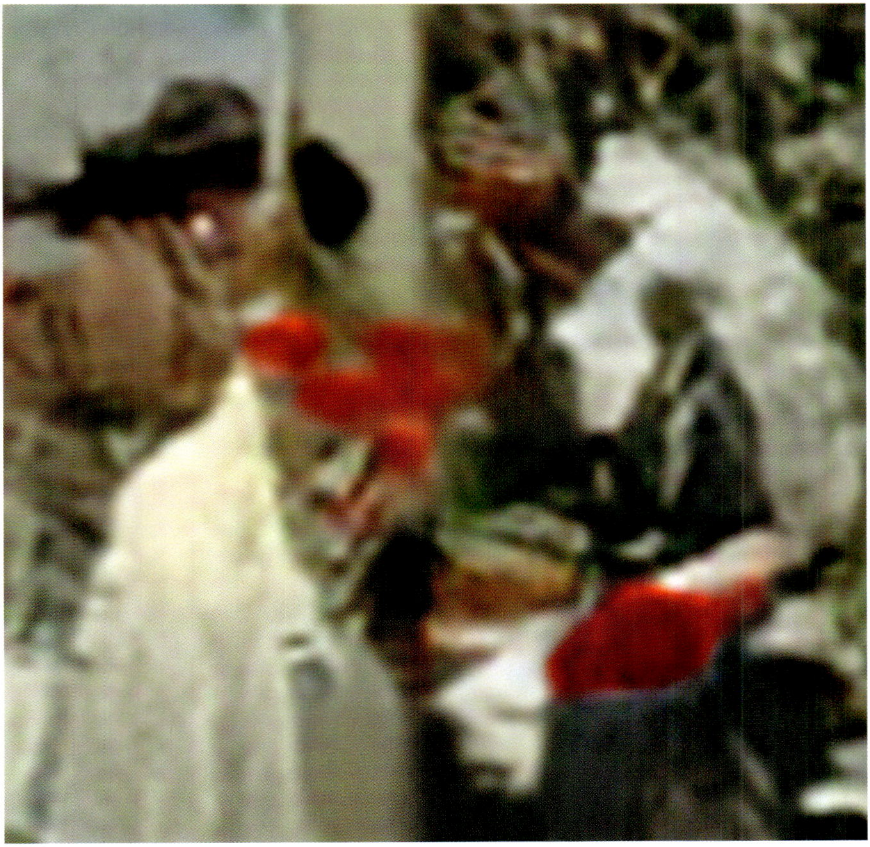

Untitled Film Stills 2.10 2019
C-print
11.8 × 11.8 in / 30 × 30 cm

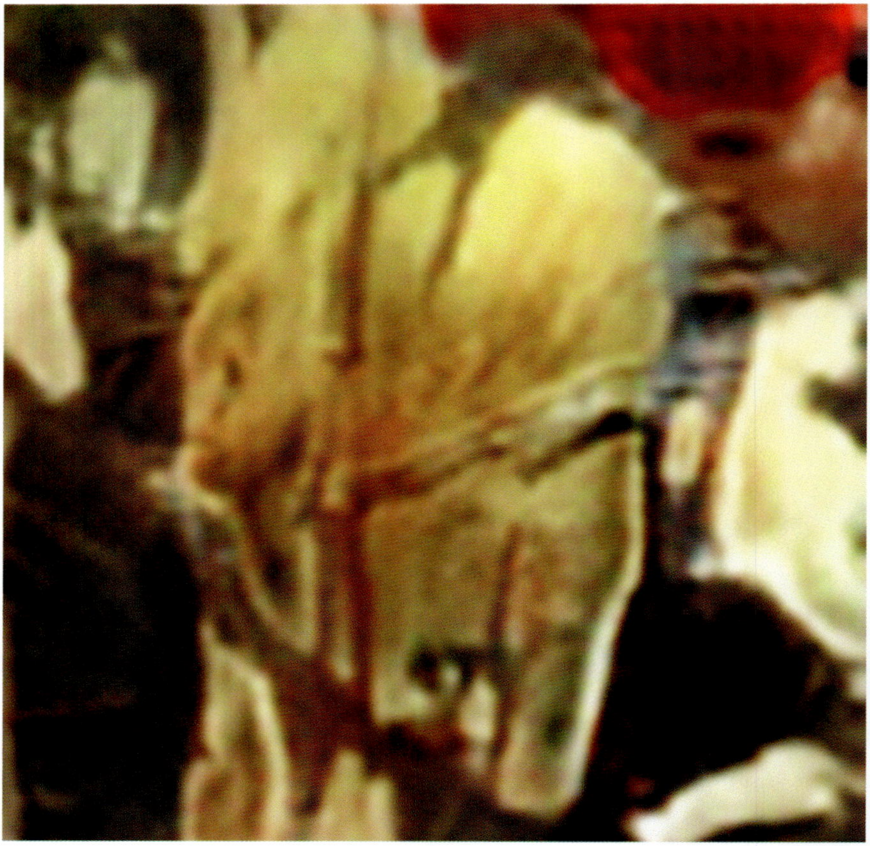

Untitled Film Stills 2.11 2019
C-print
11.8 × 11.8 in / 30 × 30 cm

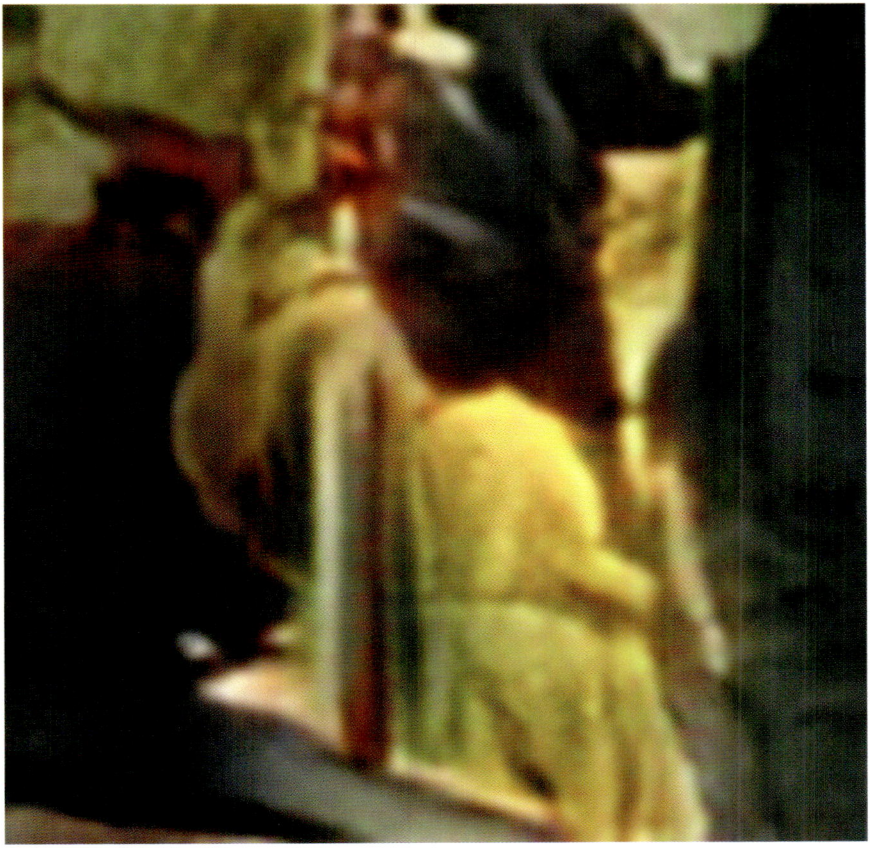

Untitled Film Stills 2.12 2019
C-print
11.8 × 11.8 in / 30 × 30 cm

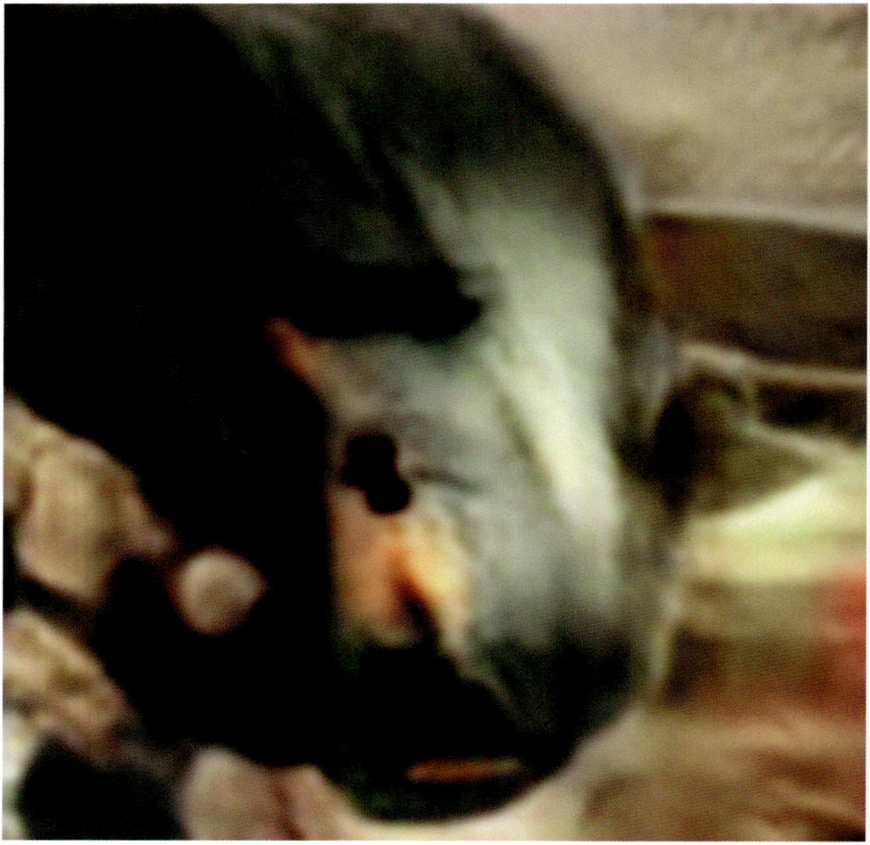

Untitled Film Stills 2.13 2019
C-print
11.8 × 11.8 in / 30 × 30 cm

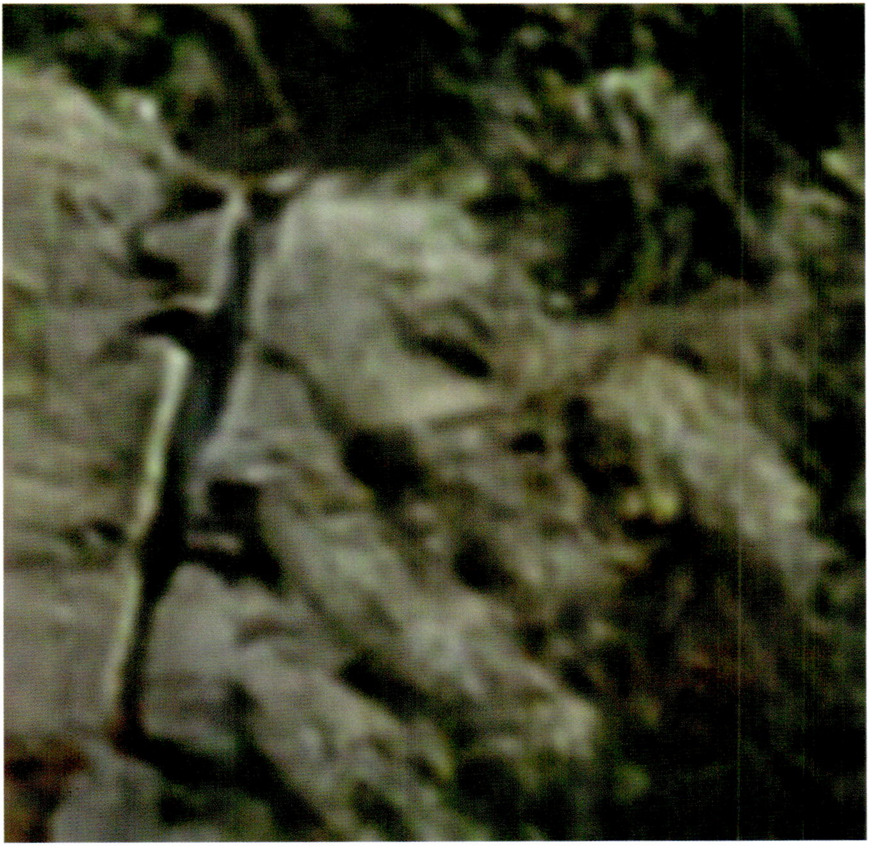

Untitled Film Stills 2.14 2019
C-print
11.8 × 11.8 in / 30 × 30 cm

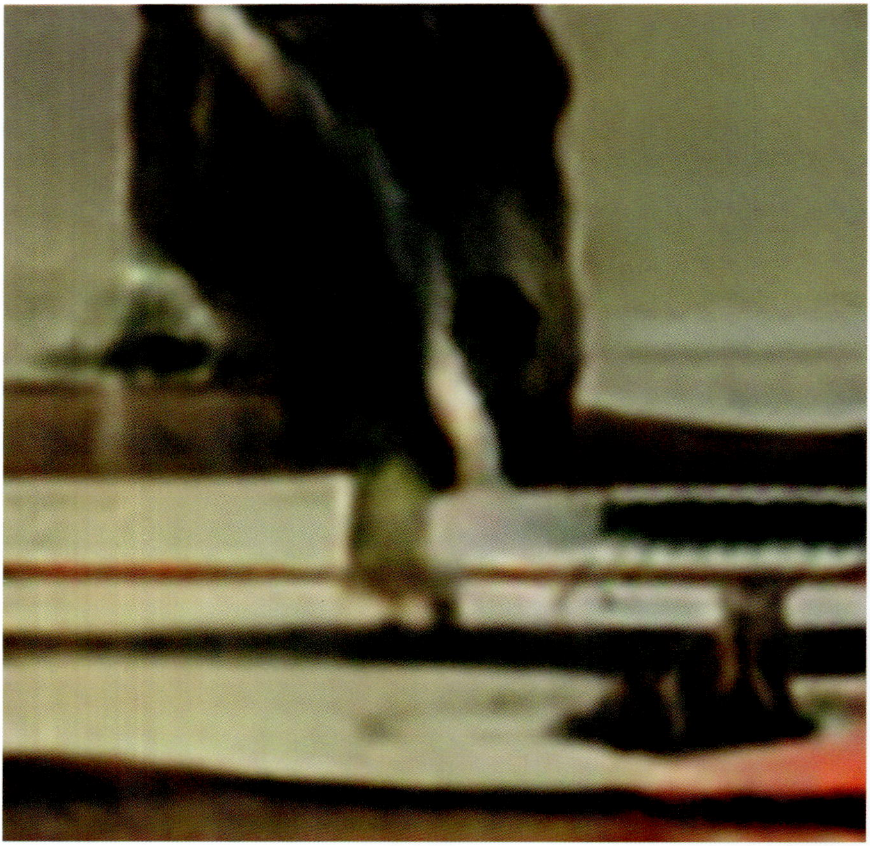

Untitled Film Stills 2.15 2019
C-print
11.8 × 11.8 in / 30 × 30 cm

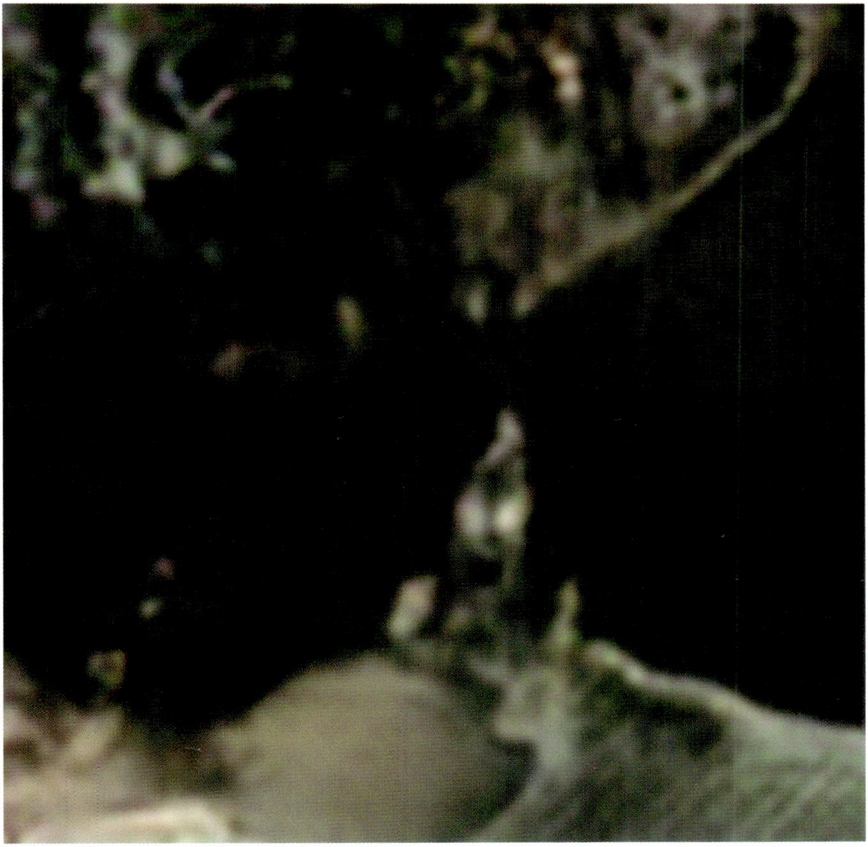

Untitled Film Stills 2.16 2019
C-print
11.8 × 11.8 in / 30 × 30 cm

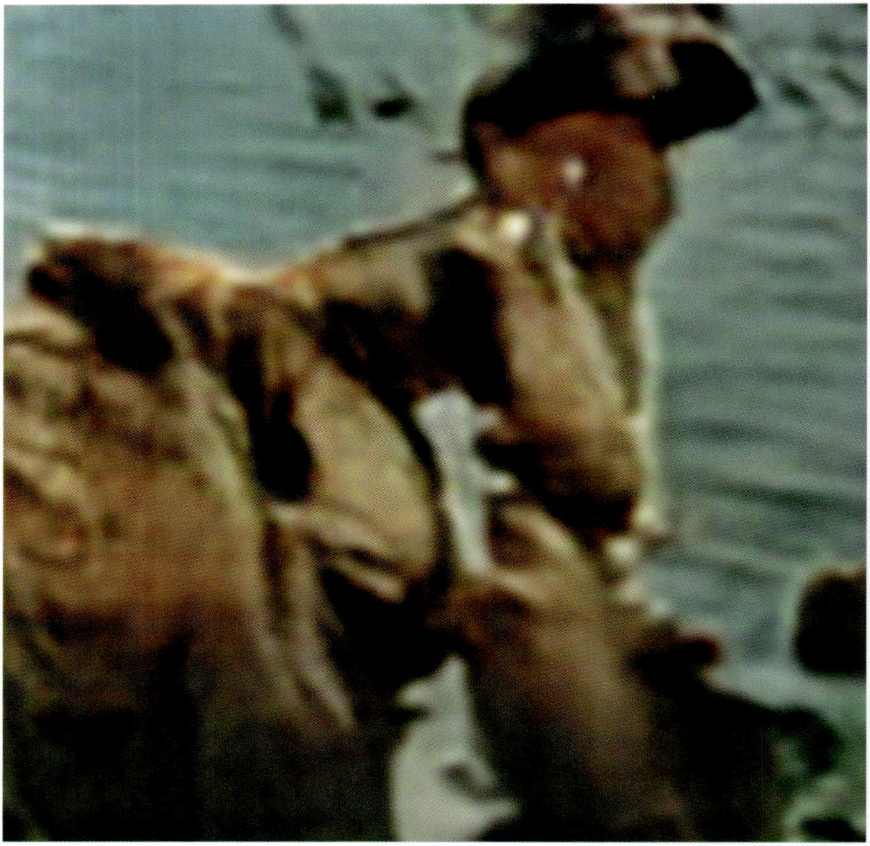

Untitled Film Stills 2.17 2019
C-print
11.8 × 11.8 in / 30 × 30 cm

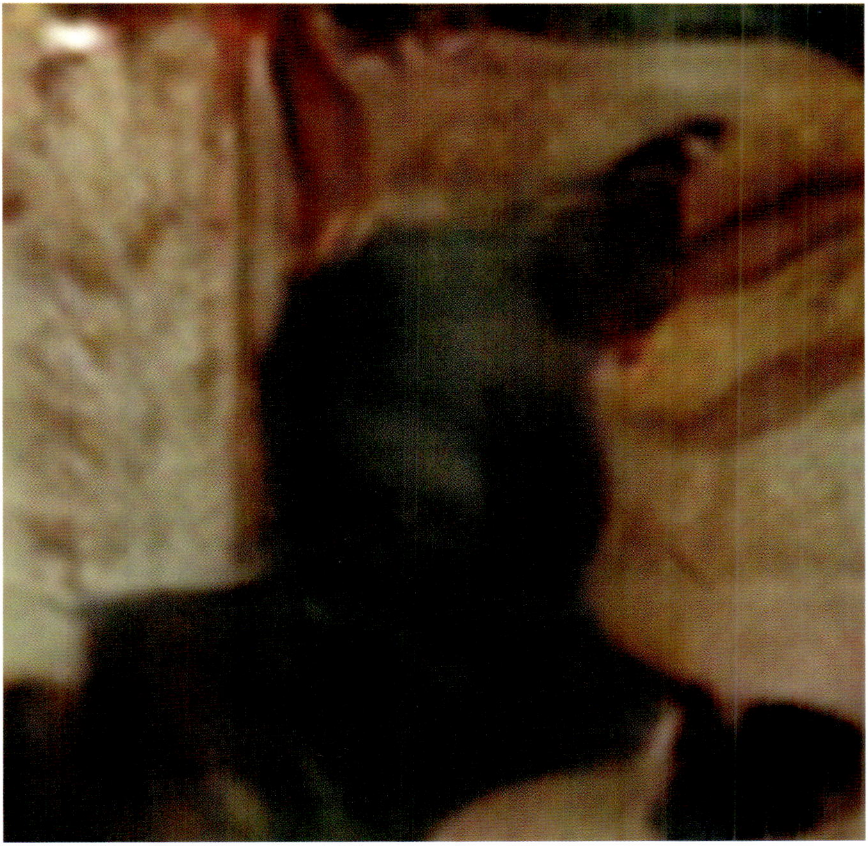

Untitled Film Stills 2.18 2019
C-print
11.8 × 11.8 in / 30 × 30 cm

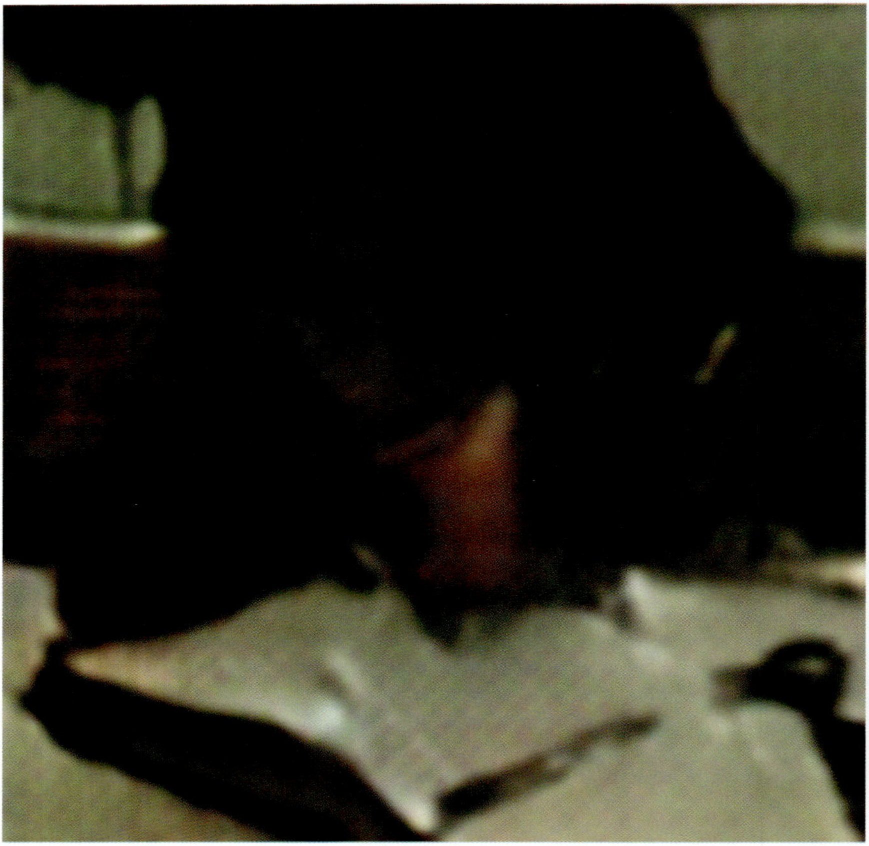

Untitled Film Stills 2.19 2019
C-print
11.8 × 11.8 in / 30 × 30 cm

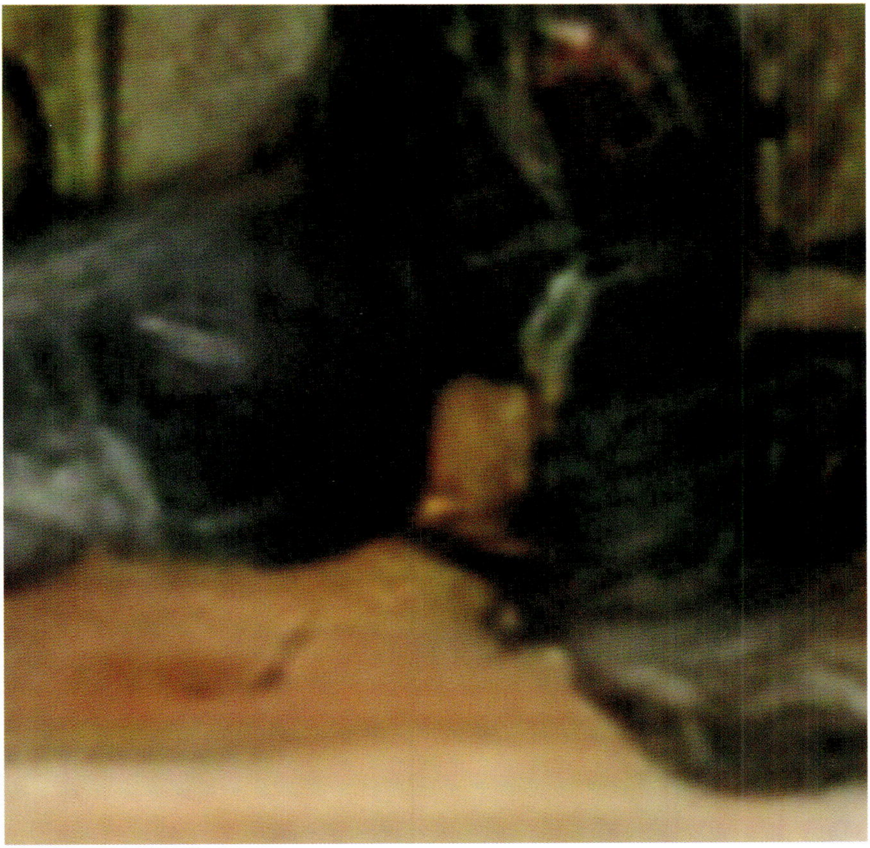

Untitled Film Stills 2.20 2019
C-print
11.8 × 11.8 in / 30 × 30 cm

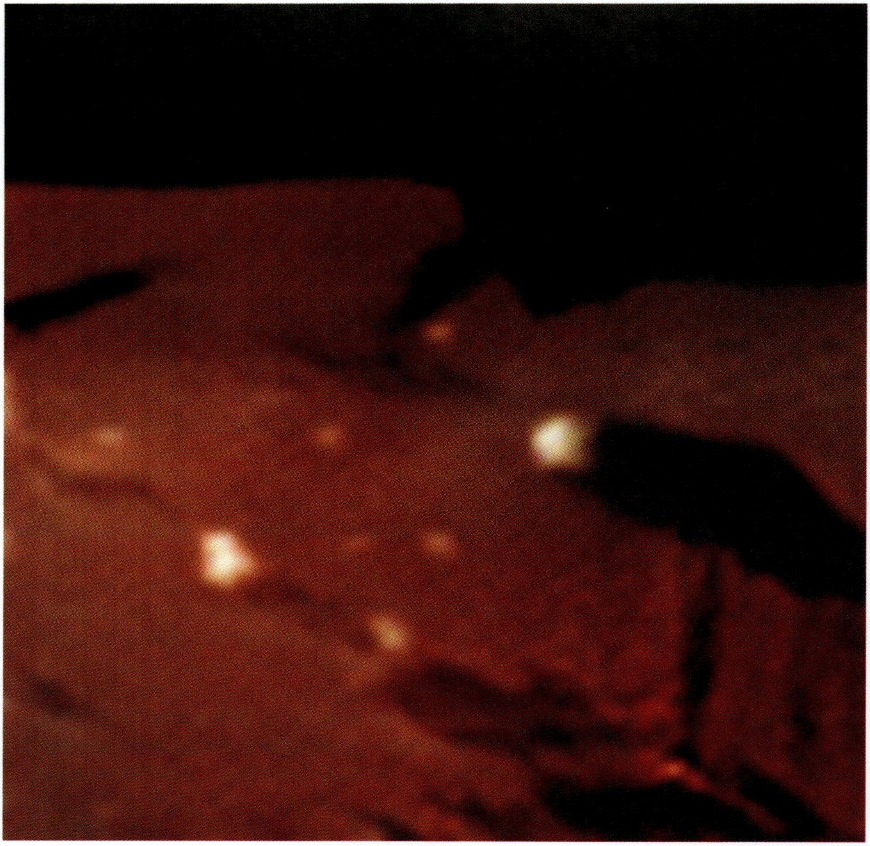

Untitled Film Stills 2.21 2019
C-print
11.8 × 11.8 in / 30 × 30 cm

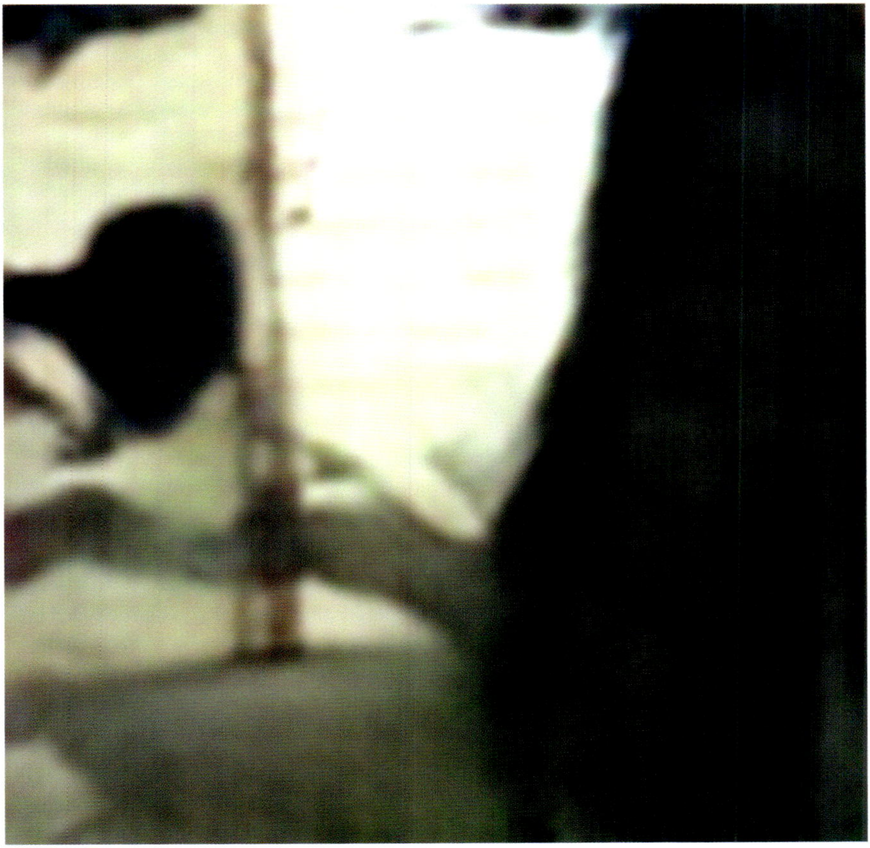

Untitled Film Stills 2.22 2019
C-print
11.8 × 11.8 in / 30 × 30 cm

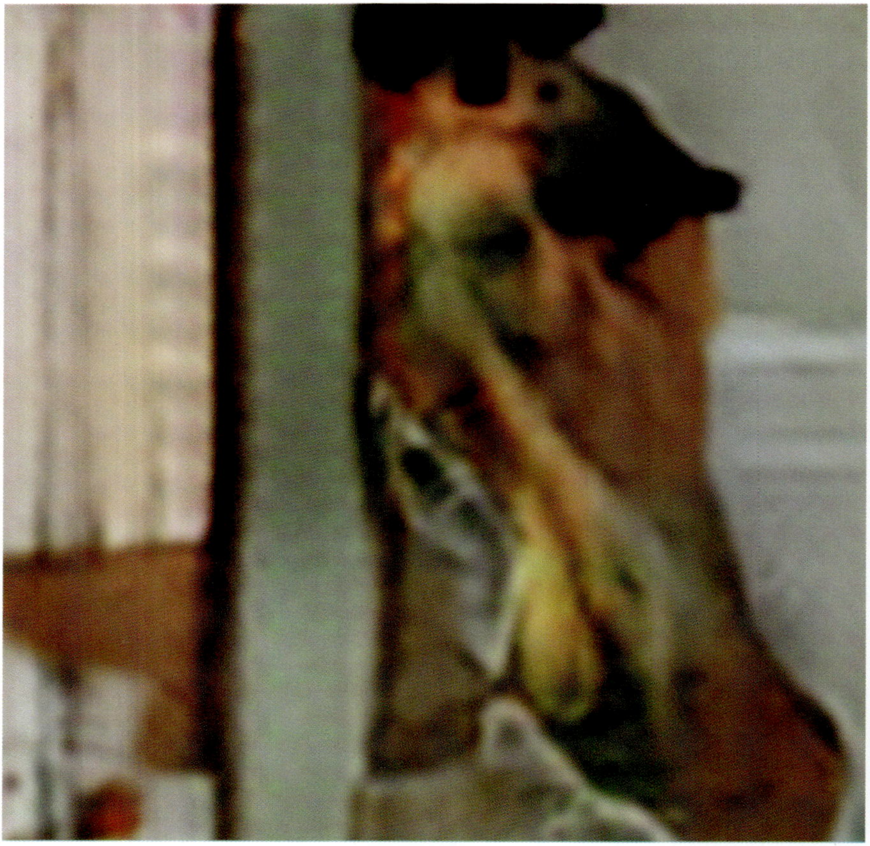

Untitled Film Stills 2.23 2019
C-print
11.8 × 11.8 in / 30 × 30 cm

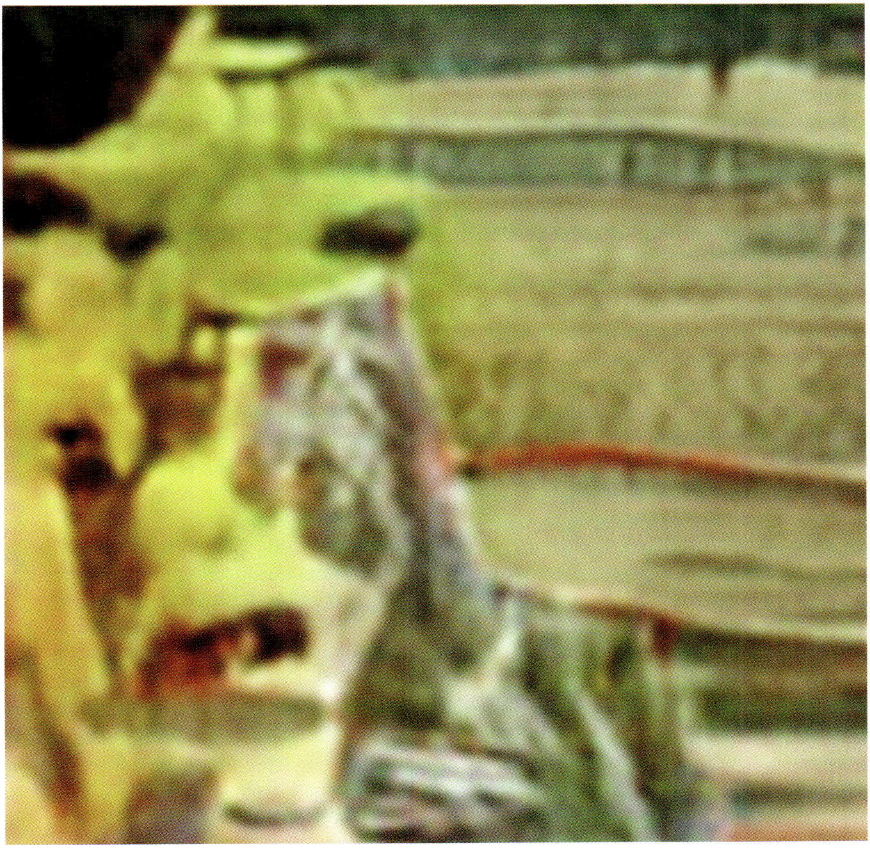

Untitled Film Stills 2.24 2019
C-print
11.8 × 11.8 in / 30 × 30 cm

Earthly Delights

The *Earthly Delights* series is a collection of software works of indefinite duration that are created from a set of fixed frames and controlled live by custom software. The software edits the frames into different sequences and presents them with changing speeds and rhythms. The frames are generated with machine learning software trained on scans of vegetation collected in Colorado. The title of these works are derived from Stan Brakhage's short film *The Garden of Earthly Delights* (1981) made from raw plant materials from the same region. These are nonlinear digital videos unconstrained from a fixed frame rate and predetermined frame sequence.

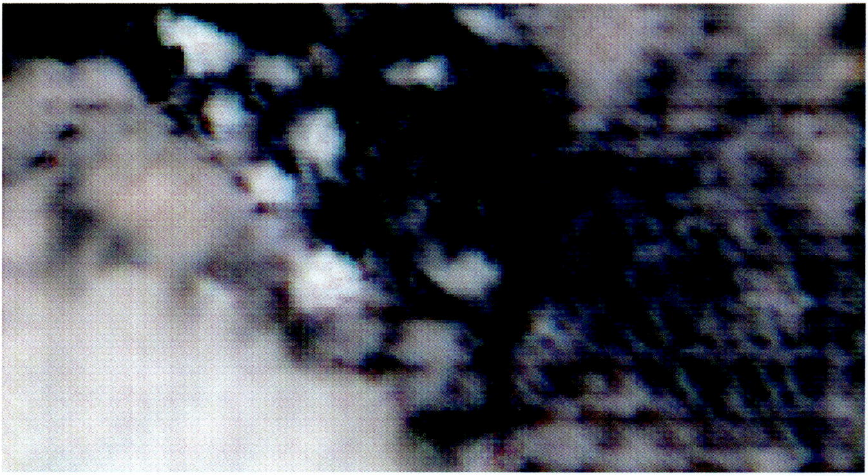

Earthly Delights 1.1 2019
Custom software (color, silent), computer, screen or projector
Dimensions variable, horizontal.

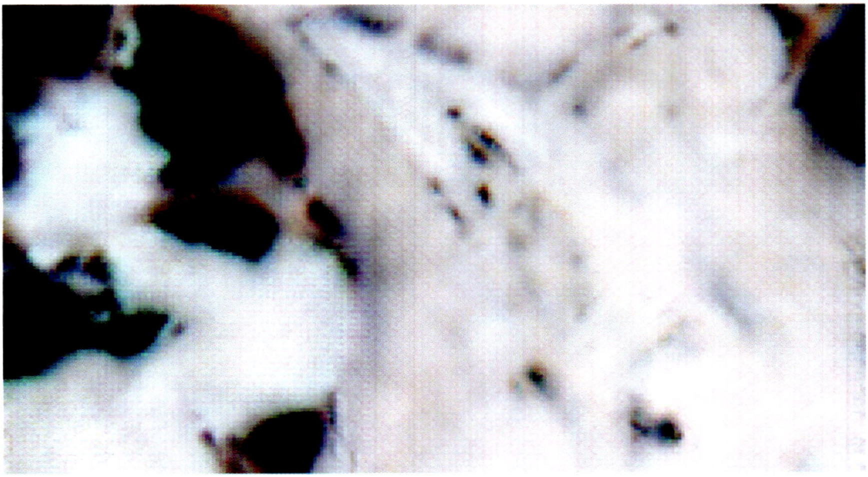

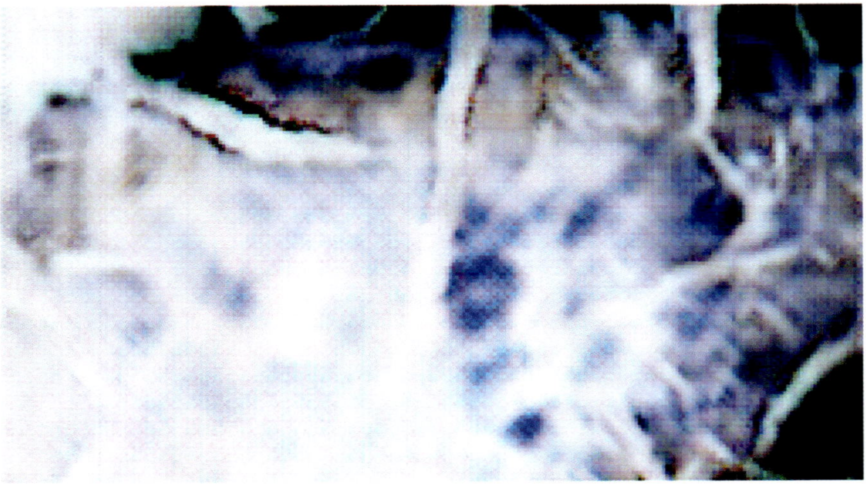

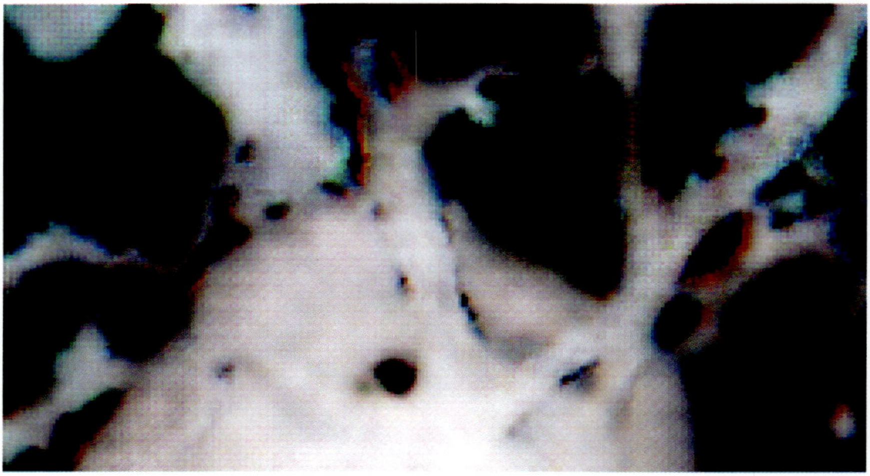

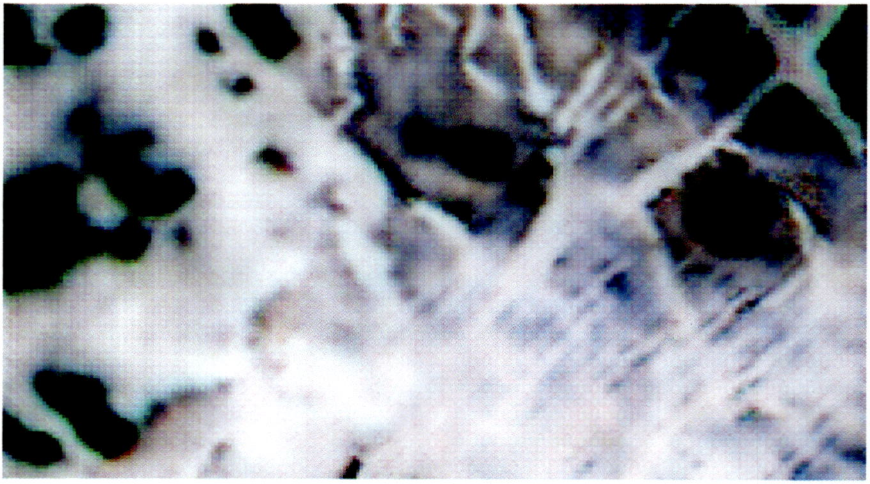

Earthly Delights 1.2 2019
Custom software (color, silent), computer, screen or projector
Dimensions variable, horizontal.

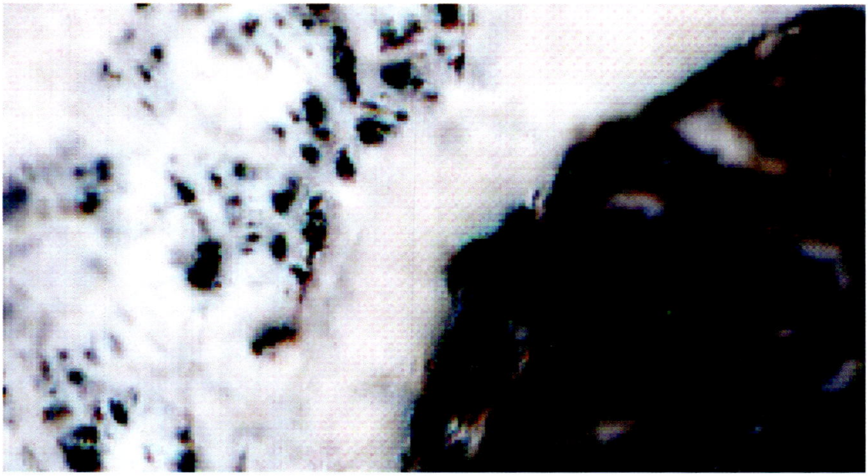

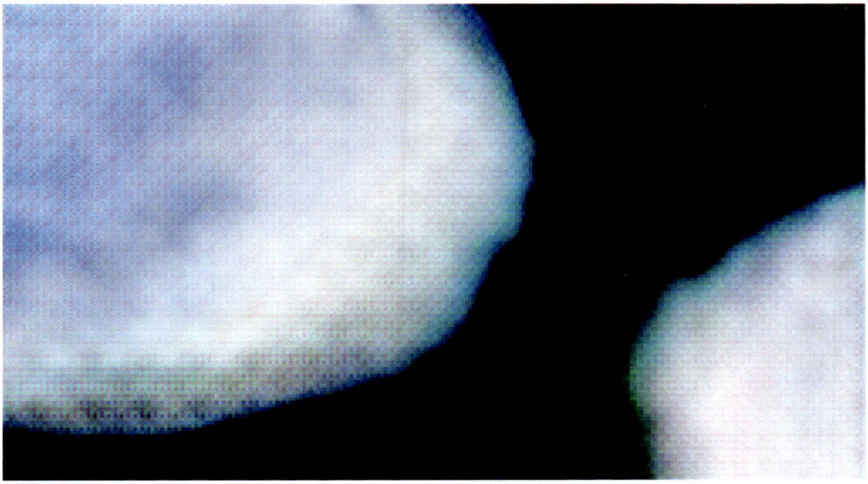

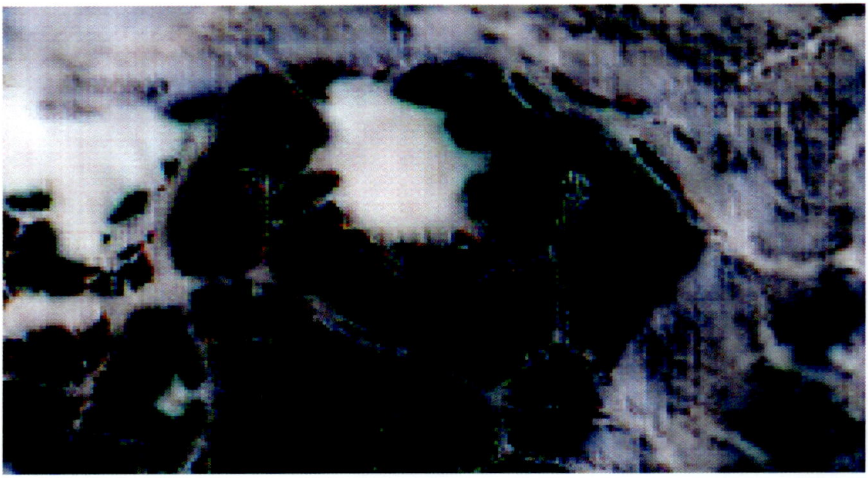

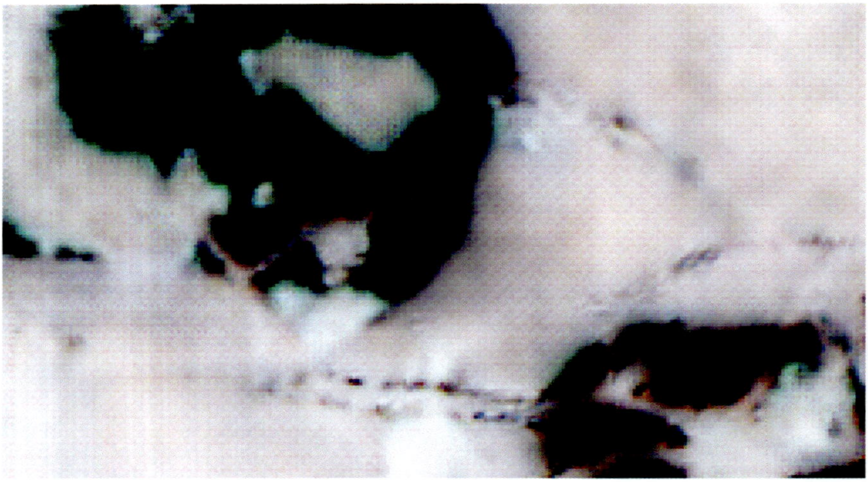

Earthly Delights 1.3 2019
Custom software (color, silent), computer, screen or projector
Dimensions variable, horizontal.

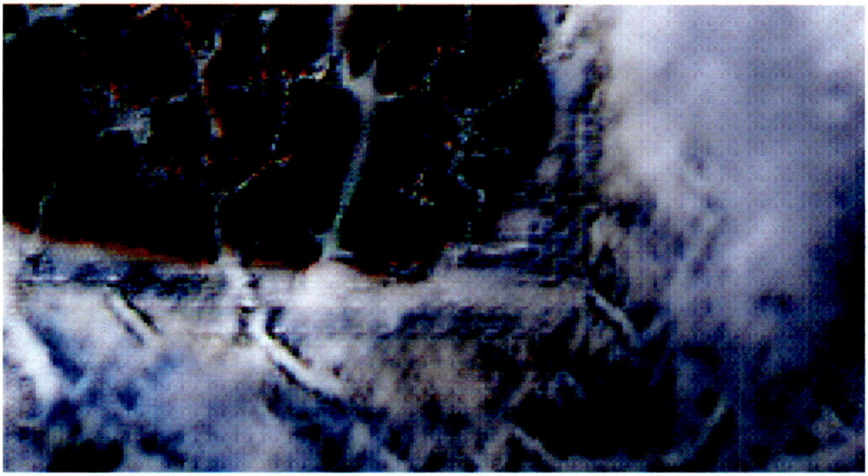

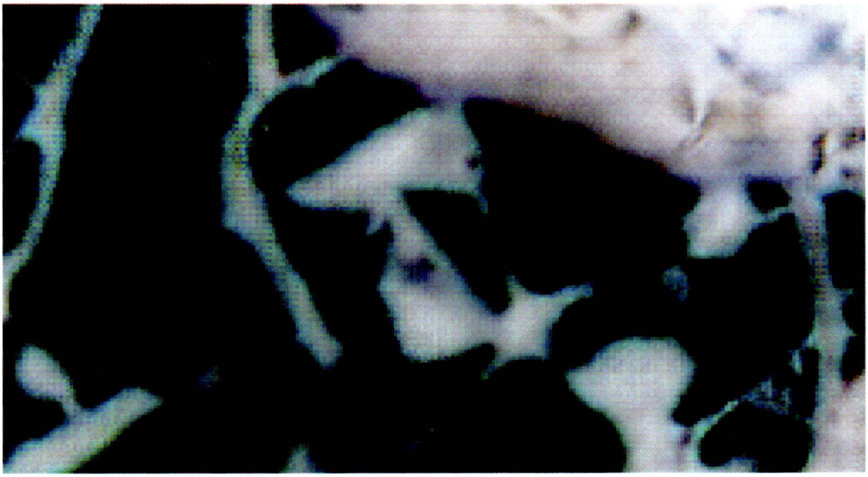

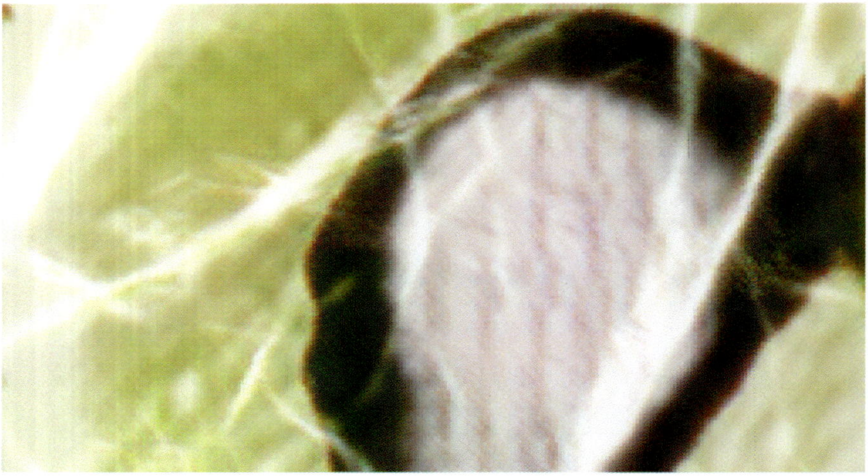

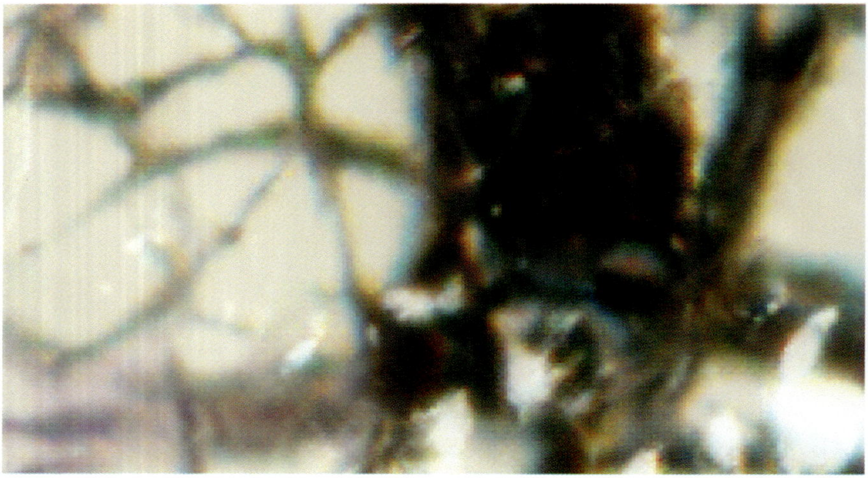

Earthly Delights 2.1 2019
Custom software (color, silent), computer, screen or projector
Dimensions variable, horizontal.

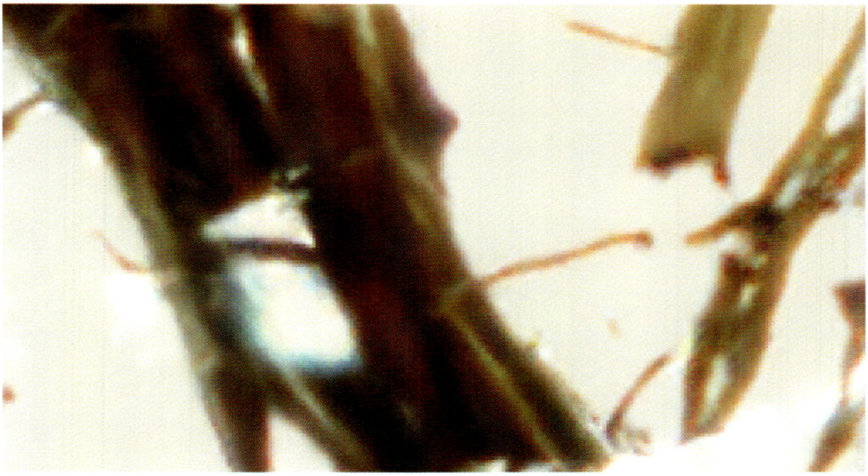

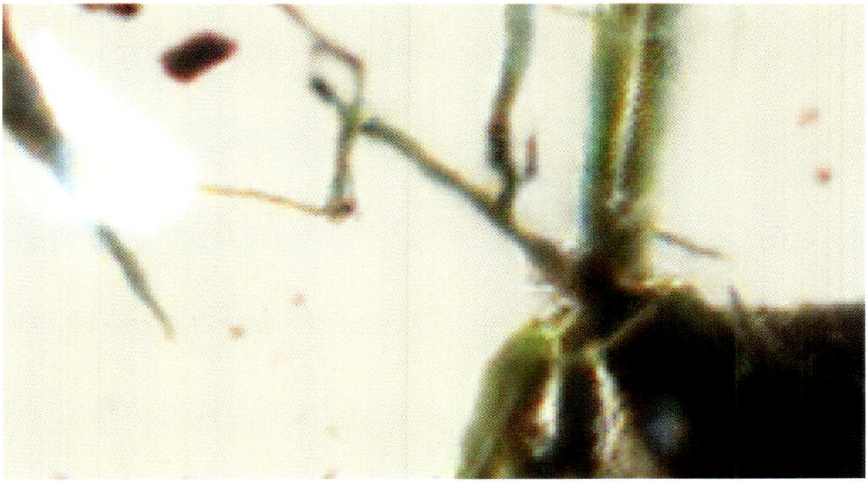

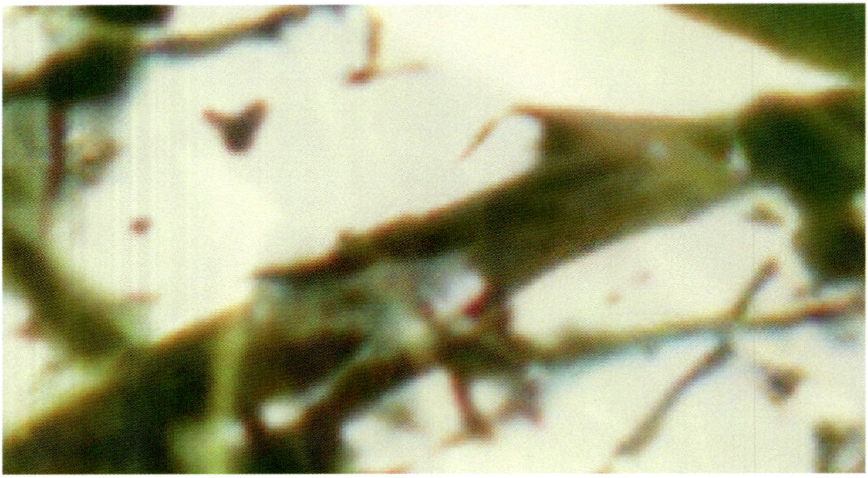

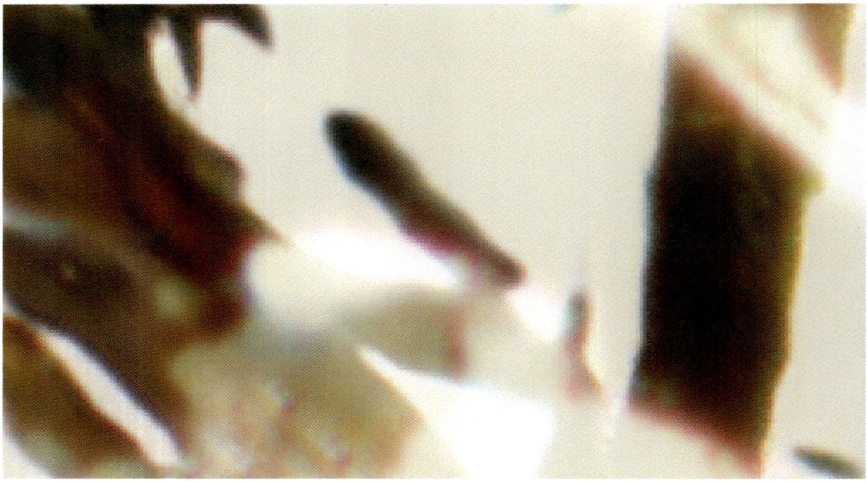

Earthly Delights 2.2 2019
Custom software (color, silent), computer, screen or projector
Dimensions variable, horizontal.

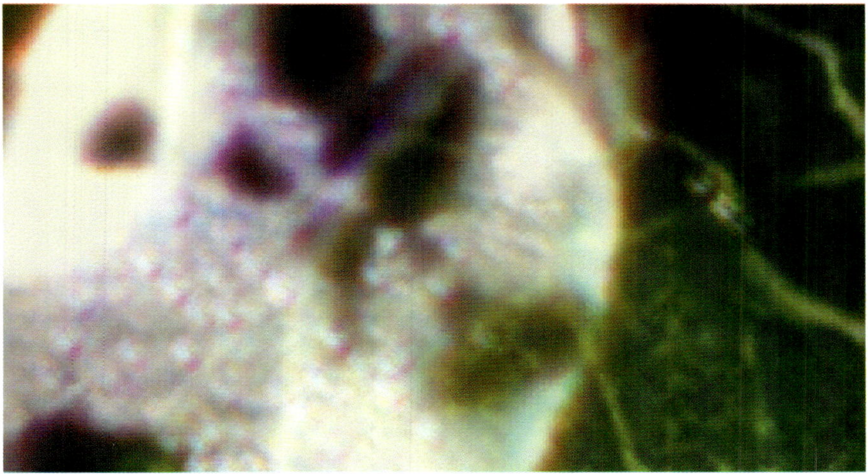

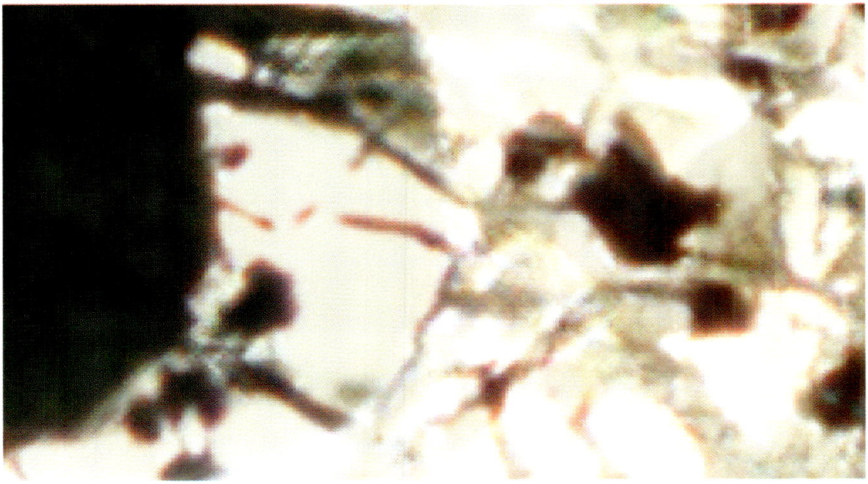

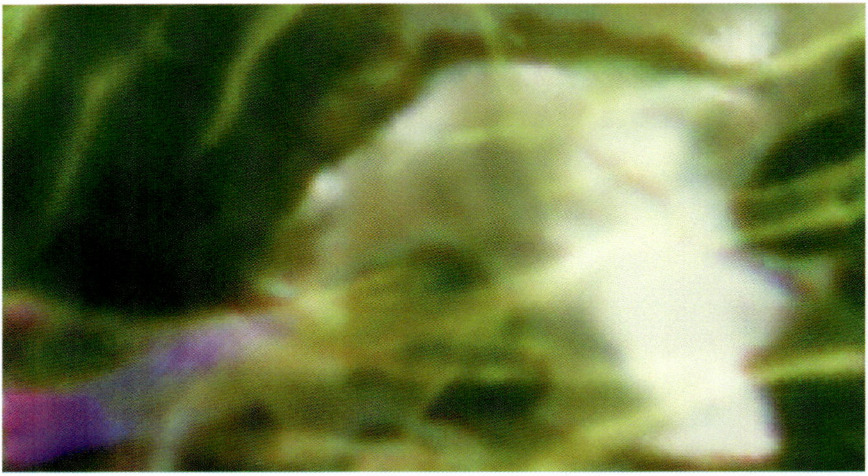

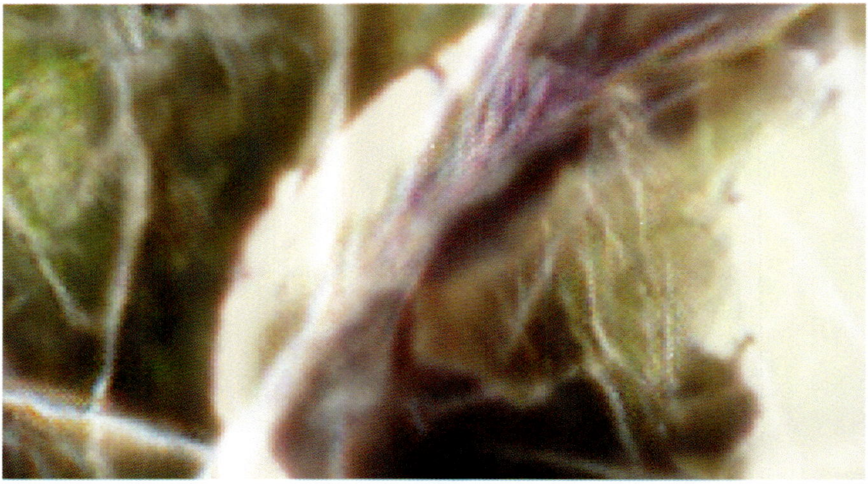

Earthly Delights 2.3 2019
Custom software (color, silent), computer, screen or projector
Dimensions variable, horizontal.

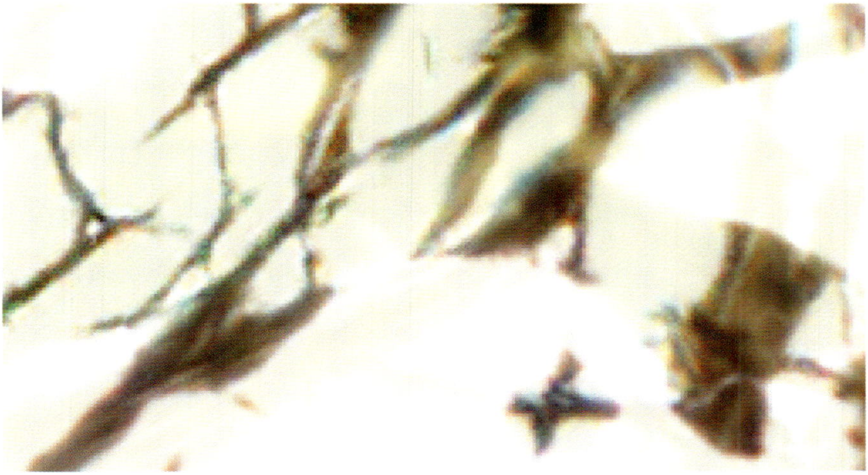

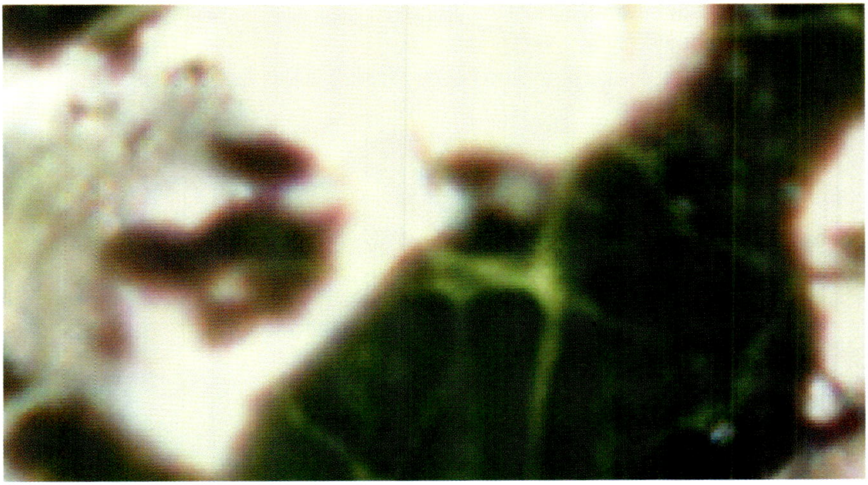

Untitled Film Stills 5.1 - 5.23

The explicit research goal of GANs is to create images that are indistinguishable from photographic images, but the aim for the *Untitled Film Stills* is to work with GANs to create uncanny images that express my personal desire for ambiguous, dreamlike images that oscillate between clarity and the ephemeral. This is controlled through the curation of the training data and the concise selection and further manipulation of the generated images. The machine learning models used for *Untitled Film Stills* are capable of producing high-fidelity, photographic images, but that's not the point of this work.

There are currently five series of *Untitled Film Stills*. Like the second series printed earlier in this text, *Untitled Film Stills 5.1 - 5.23* are images generated from a GAN model trained on over 100,000 images extracted from every individual frame of a selected film. While the second series used DCGAN, this series utilized code from the Progressive Growing of GANs research paper by Karras et al. The training images and the original outputs from the model are 512 × 512 pixels.

The first set of *Untitled Film Stills 5.1 - 5.23* were exhibited at the Share Festival in Turin in May 2019.

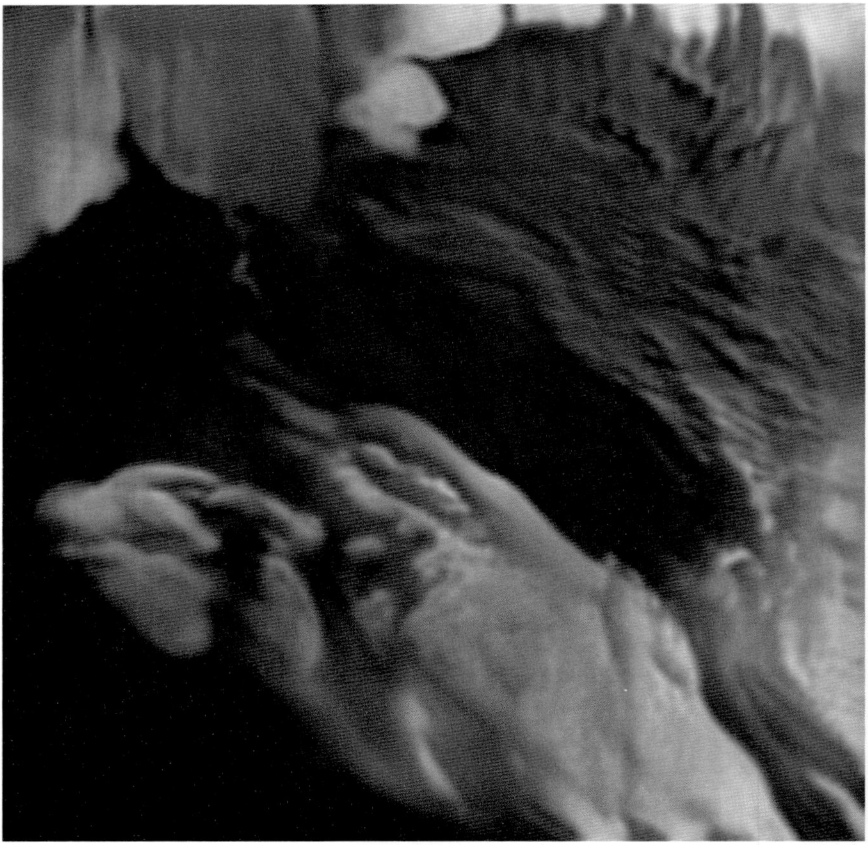

Untitled Film Stills 5.1 2019

C-print

16.5 × 16.5 in / 41.9 × 41.9 cm

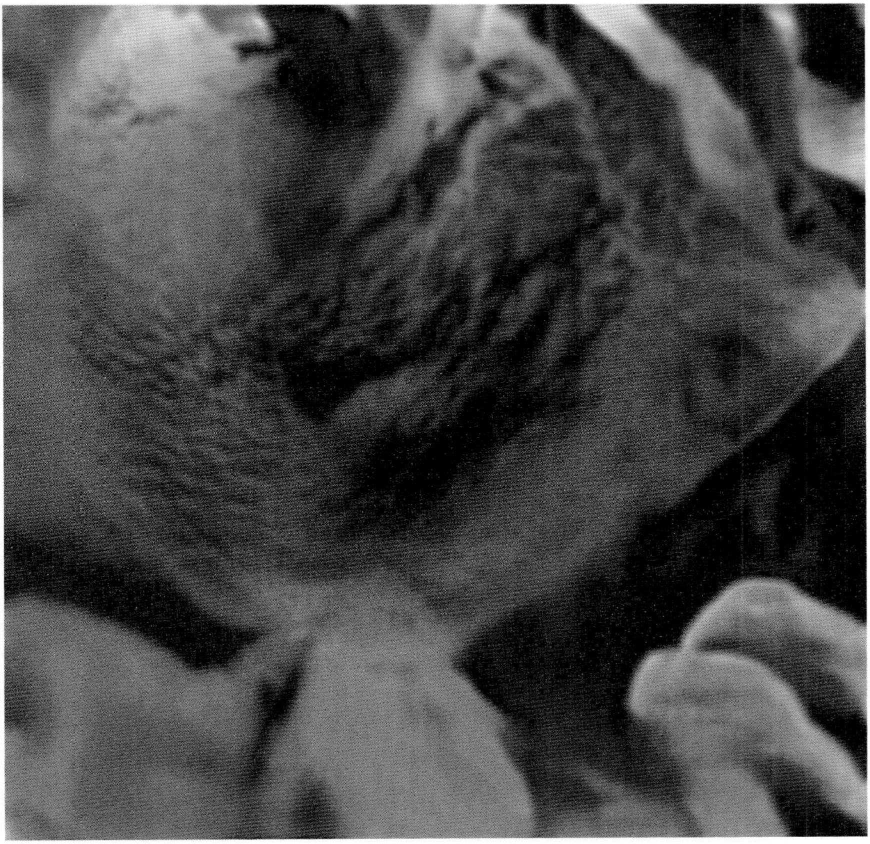

Untitled Film Stills 5. 2 2019
C-print
16.5 × 16.5 in / 41.9 × 41.9 cm

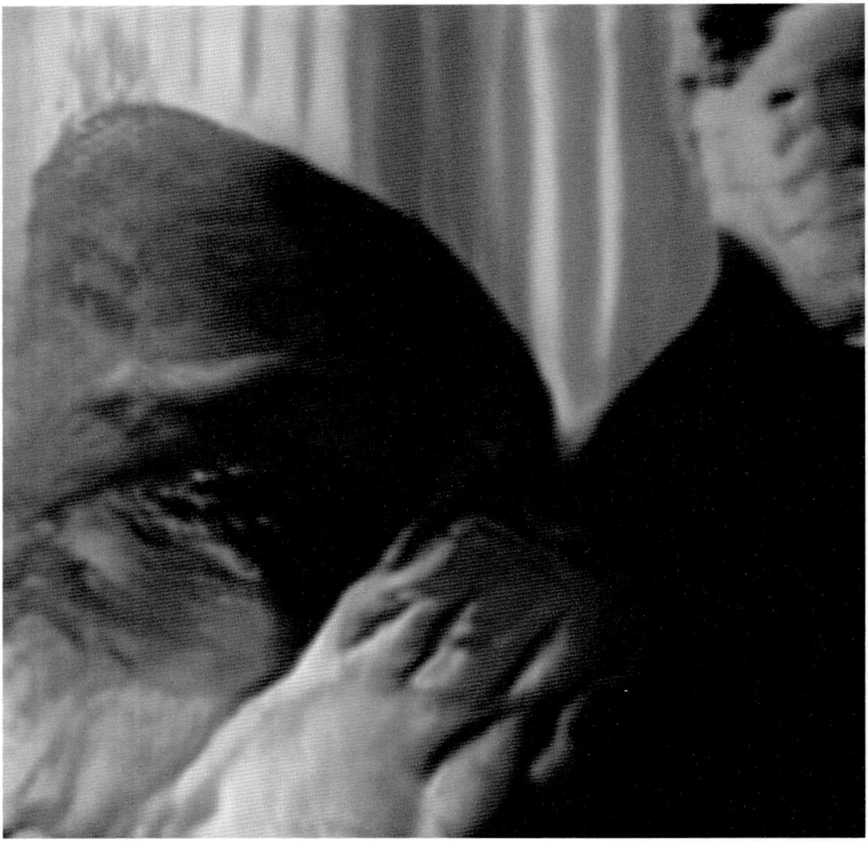

Untitled Film Still 5.3 2019
C-print
16.5 × 16.5 in / 41.9 × 41.9 cm

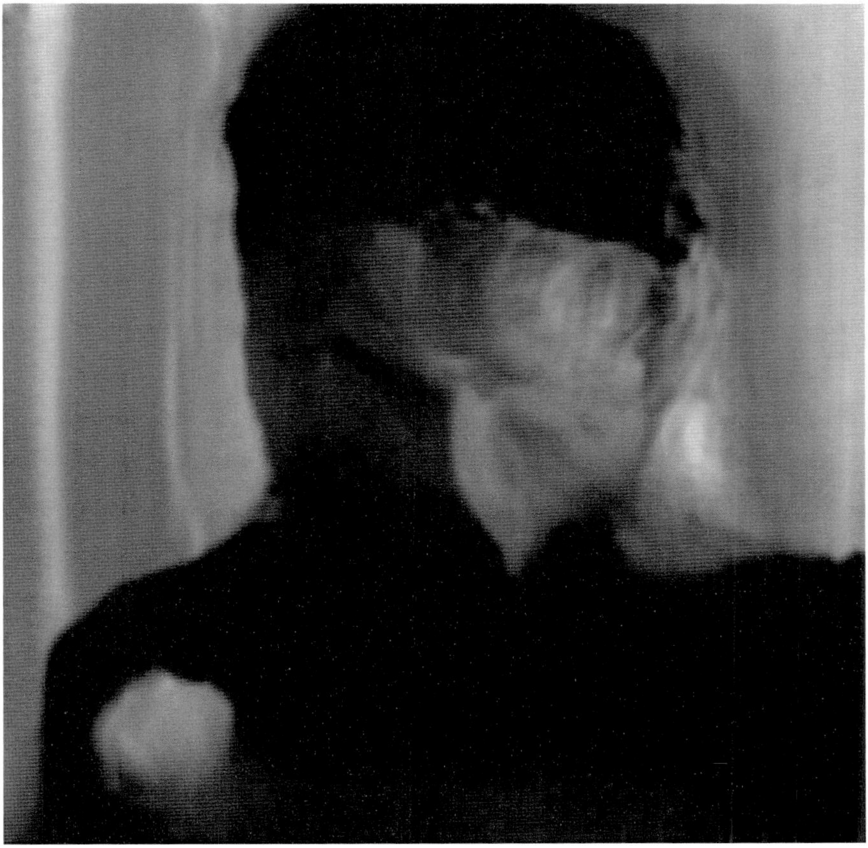

Untitled Film Still 5.4 2019
C-print
16.5 × 16.5 in / 41.9 × 41.9 cm

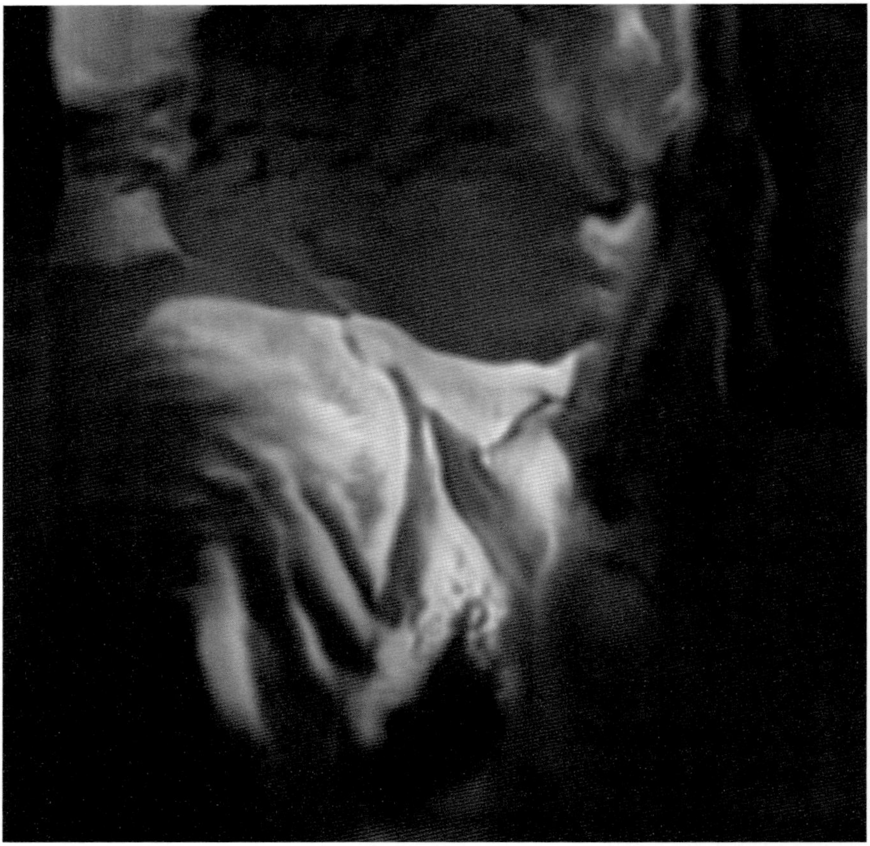

Untitled Film Still 5.5 2019
C-print
16.5 × 16.5 in / 41.9 × 41.9 cm

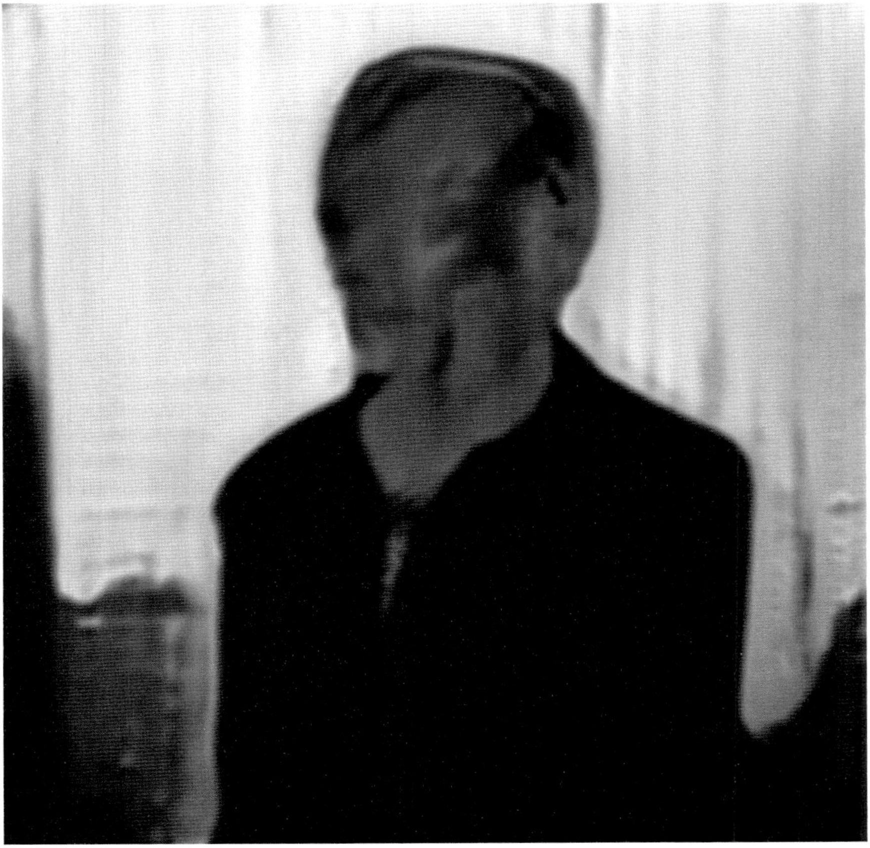

Untitled Film Still 5.6 2019
C-print
16.5 × 16.5 in / 41.9 × 41.9 cm

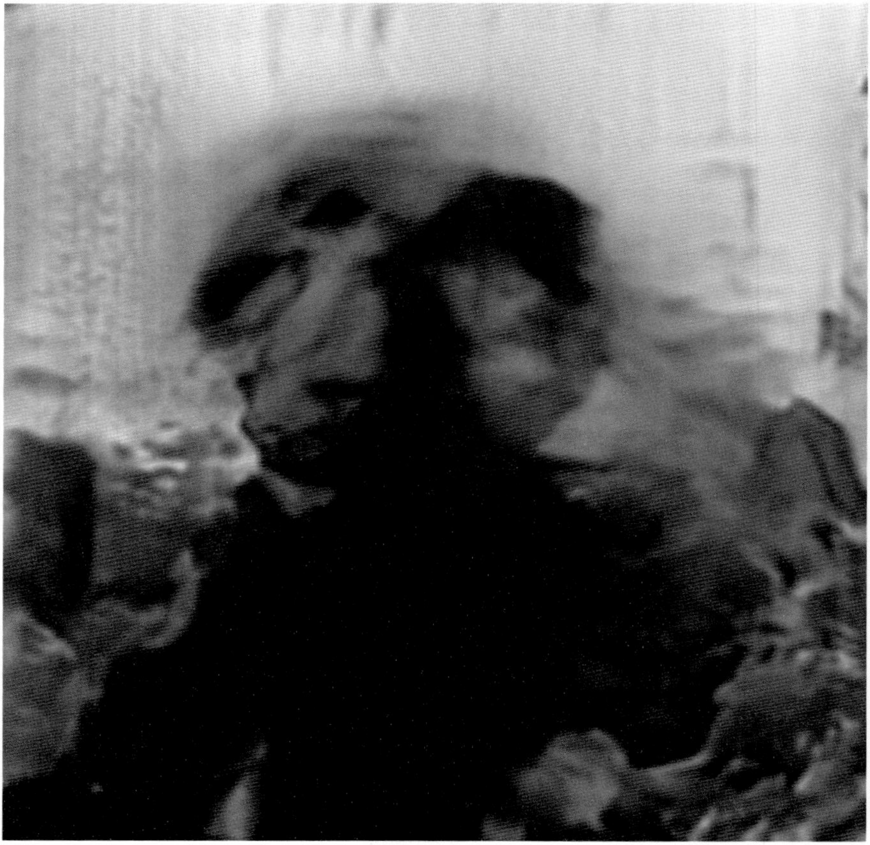

Untitled Film Still 5.7 2019
C-print
16.5 × 16.5 in / 41.9 × 41.9 cm

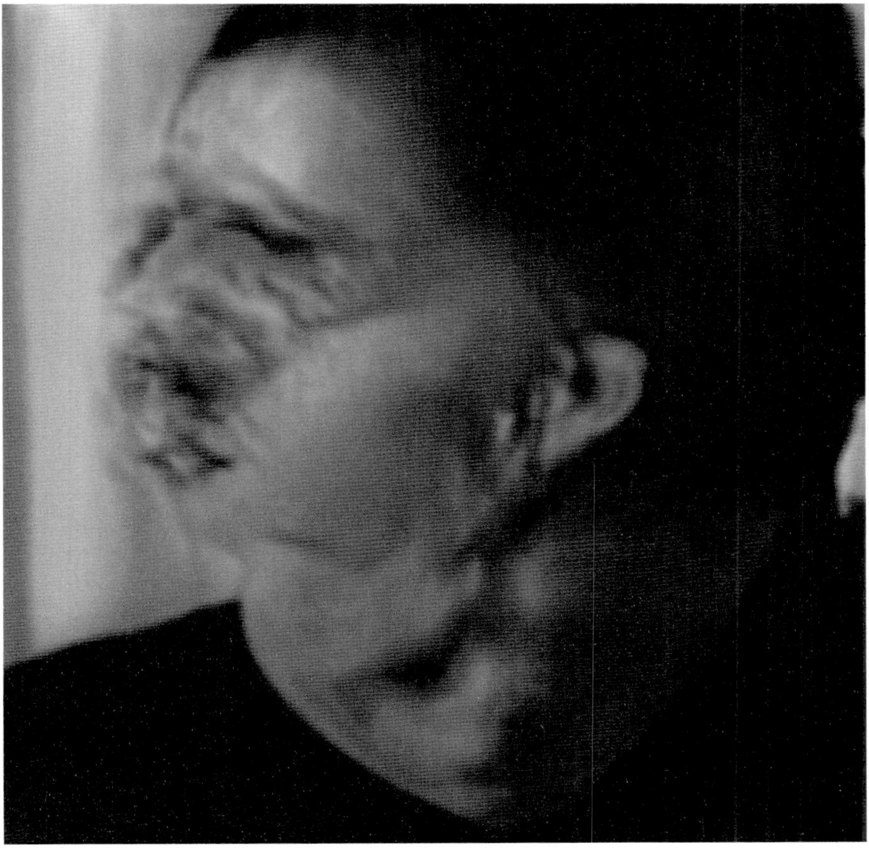

Untitled Film Still 5.8 2019
C-print
16.5 × 16.5 in / 41.9 × 41.9 cm

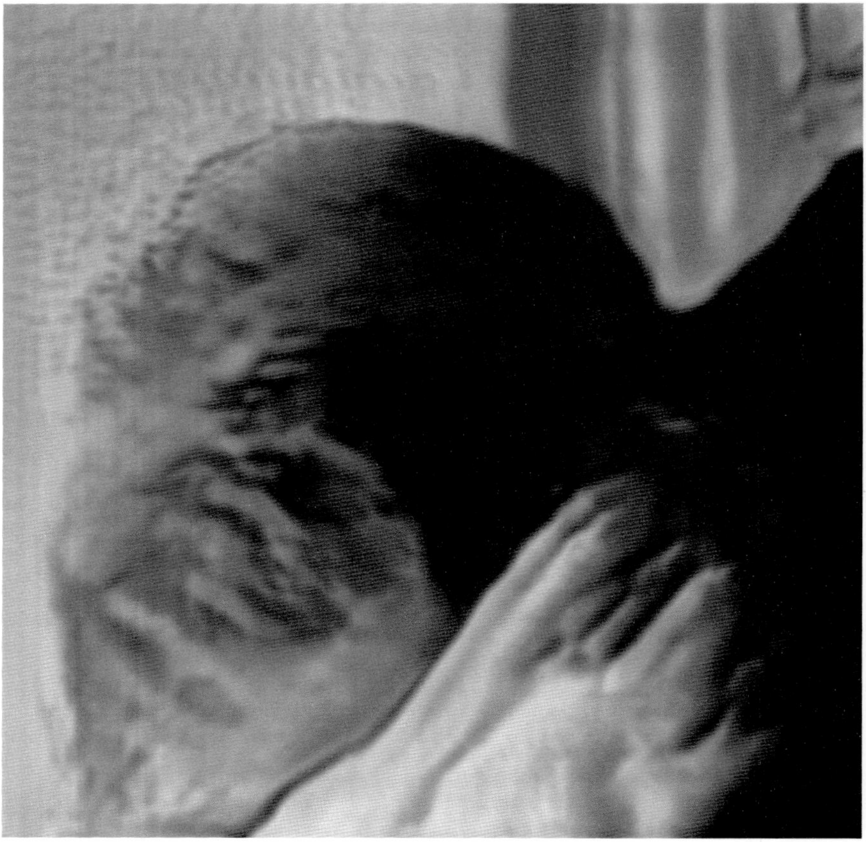

Untitled Film Still 5.9 2019
C-print
16.5 × 16.5 in / 41.9 × 41.9 cm

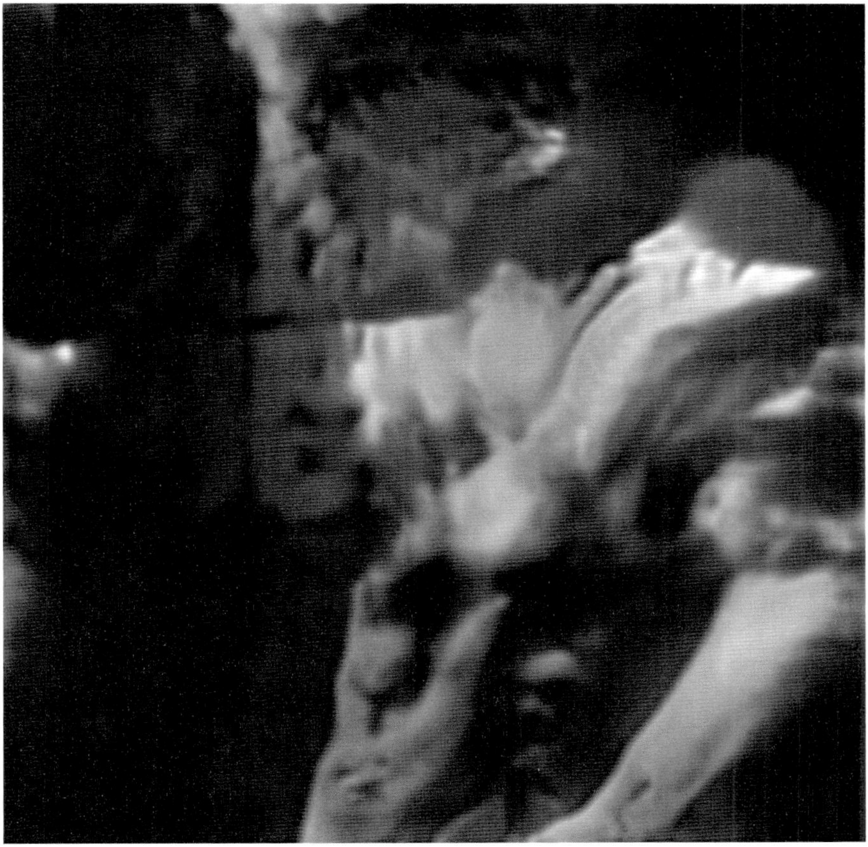

Untitled Film Still 5.10 2019
C-print
16.5 × 16.5 in / 41.9 × 41.9 cm

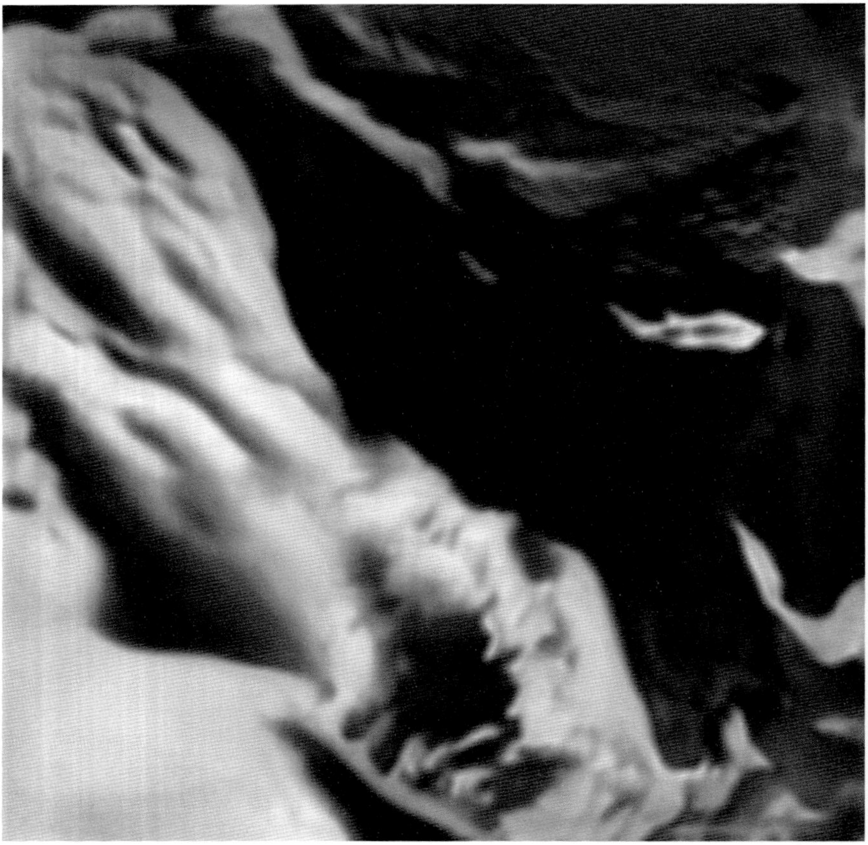

Untitled Film Still 5.11 2019
C-print
16.5 × 16.5 in / 41.9 × 41.9 cm

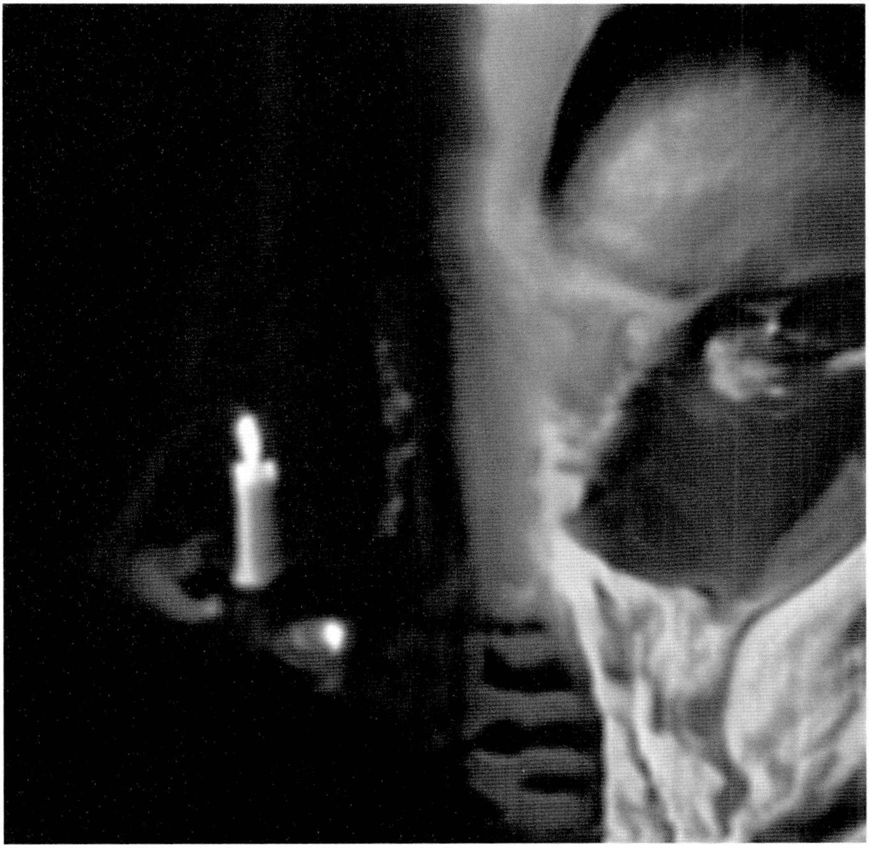

Untitled Film Still 5.12 2019
C-print
16.5 × 16.5 in / 41.9 × 41.9 cm

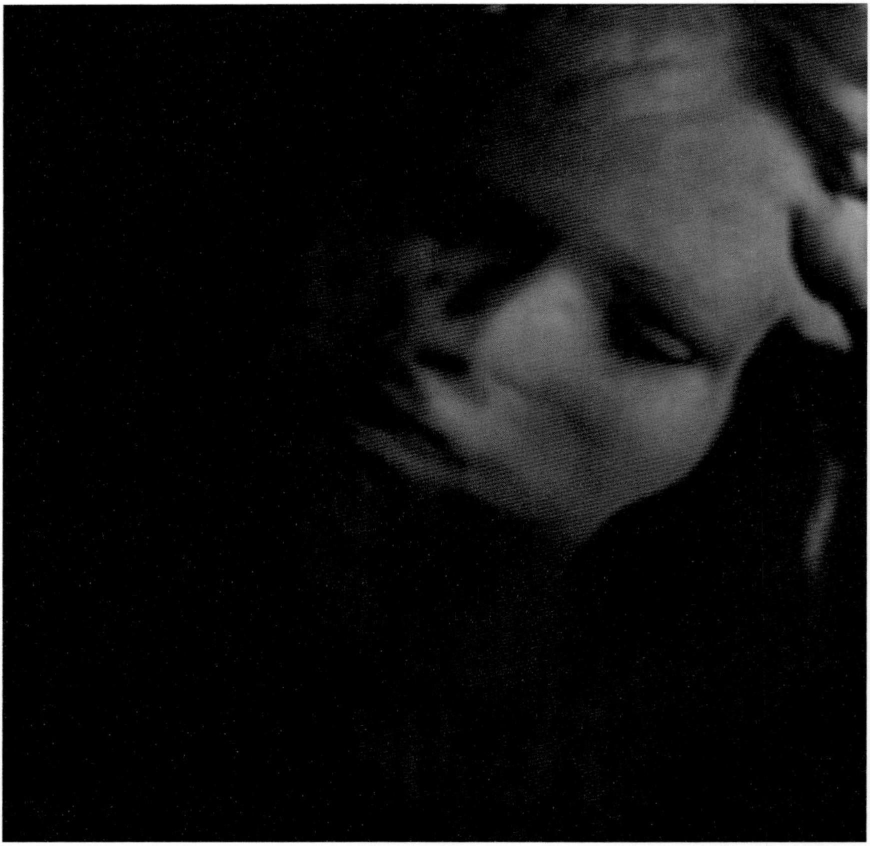

Untitled Film Still 5.13 2019
C-print
16.5 × 16.5 in / 41.9 × 41.9 cm

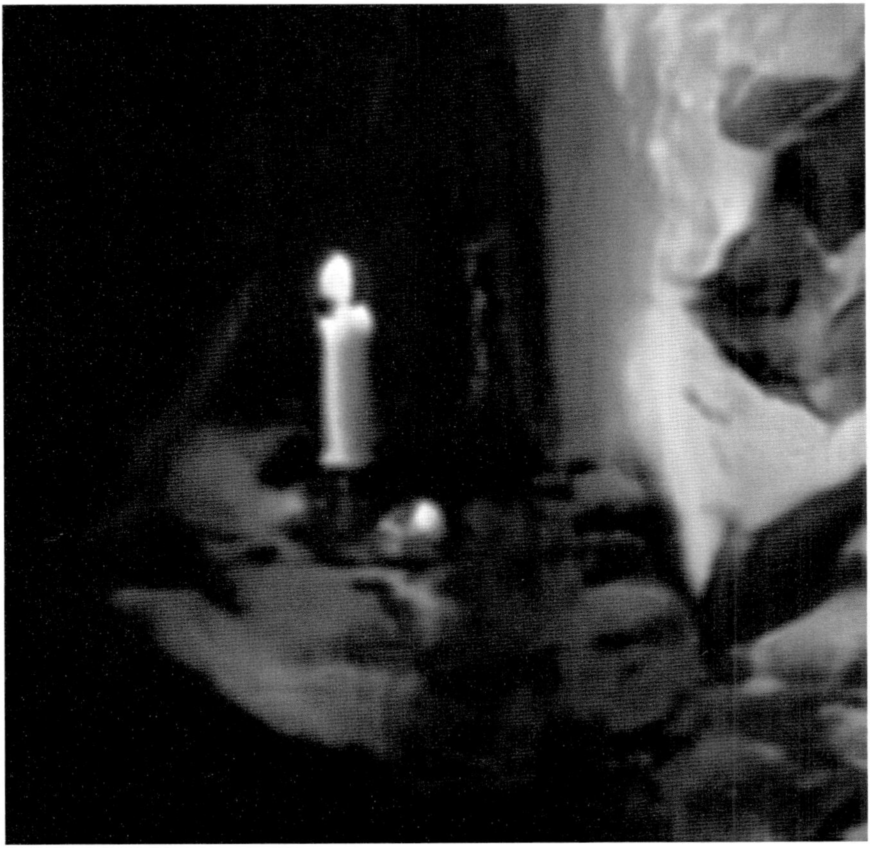

Untitled Film Still 5.14 2019
C-print
16.5 × 16.5 in / 41.9 × 41.9 cm

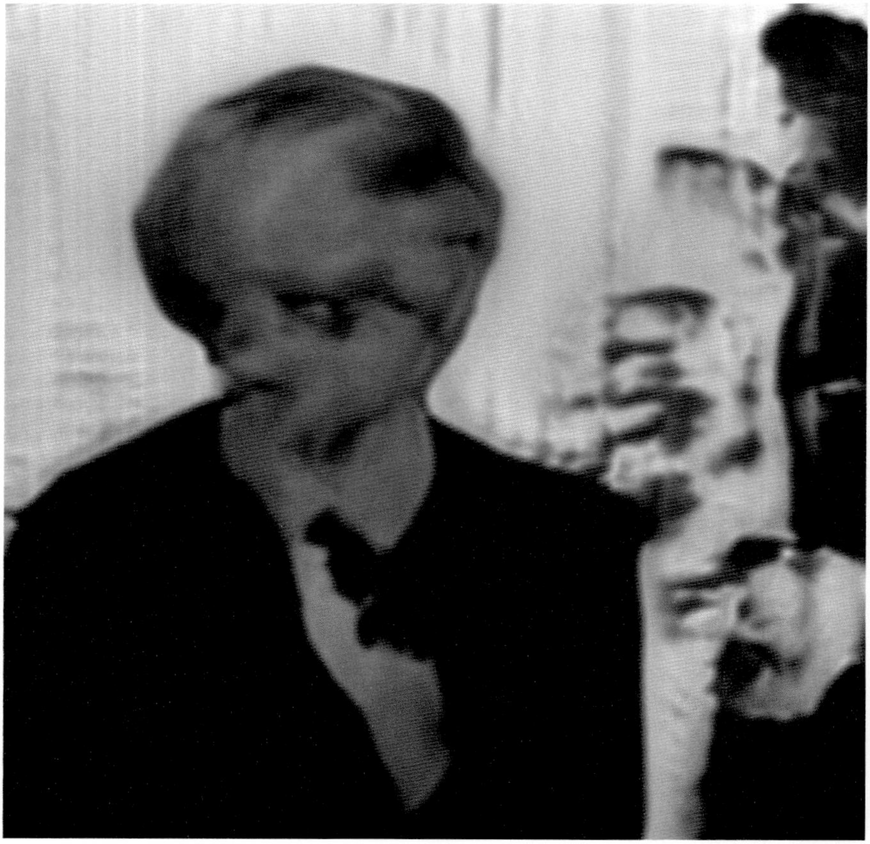

Untitled Film Still 5.15 2019
C-print
16.5 × 16.5 in / 41.9 × 41.9 cm

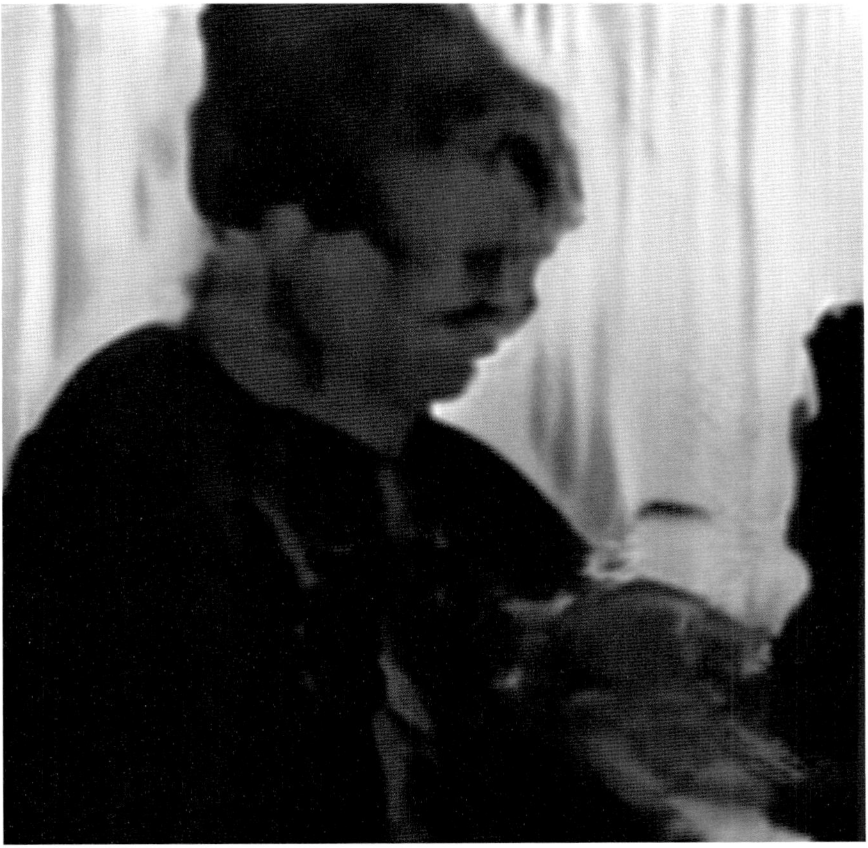

Untitled Film Still 5.16 2019
C-print
16.5 × 16.5 in / 41.9 × 41.9 cm

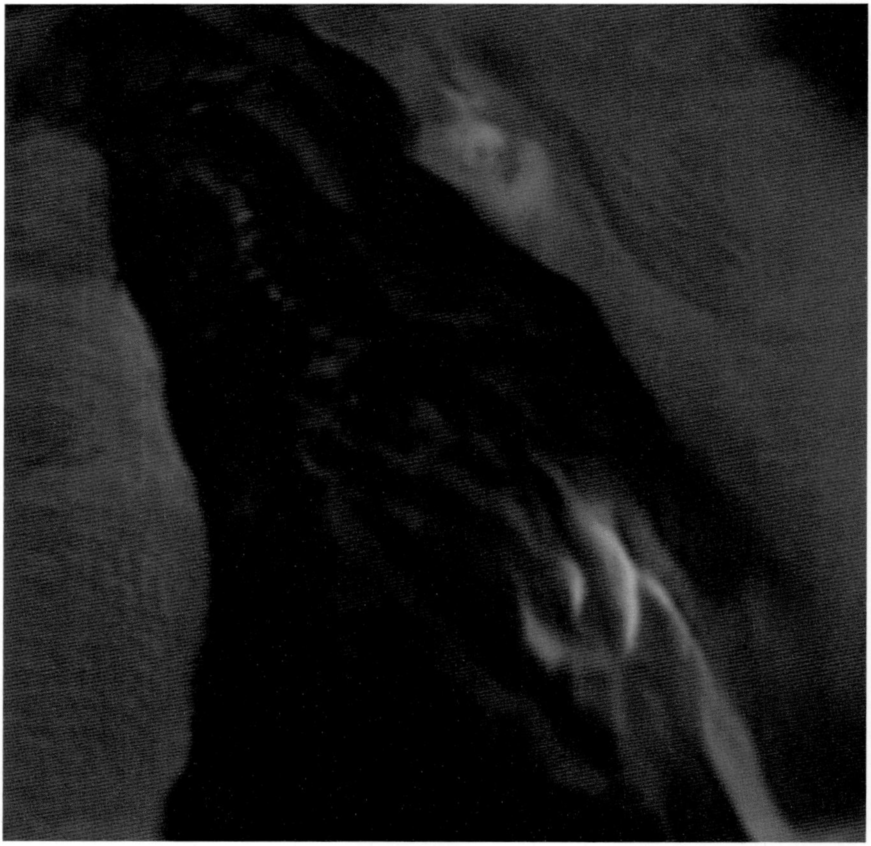

Untitled Film Still 5.17 2019
C-print
16.5 × 16.5 in / 41.9 × 41.9 cm

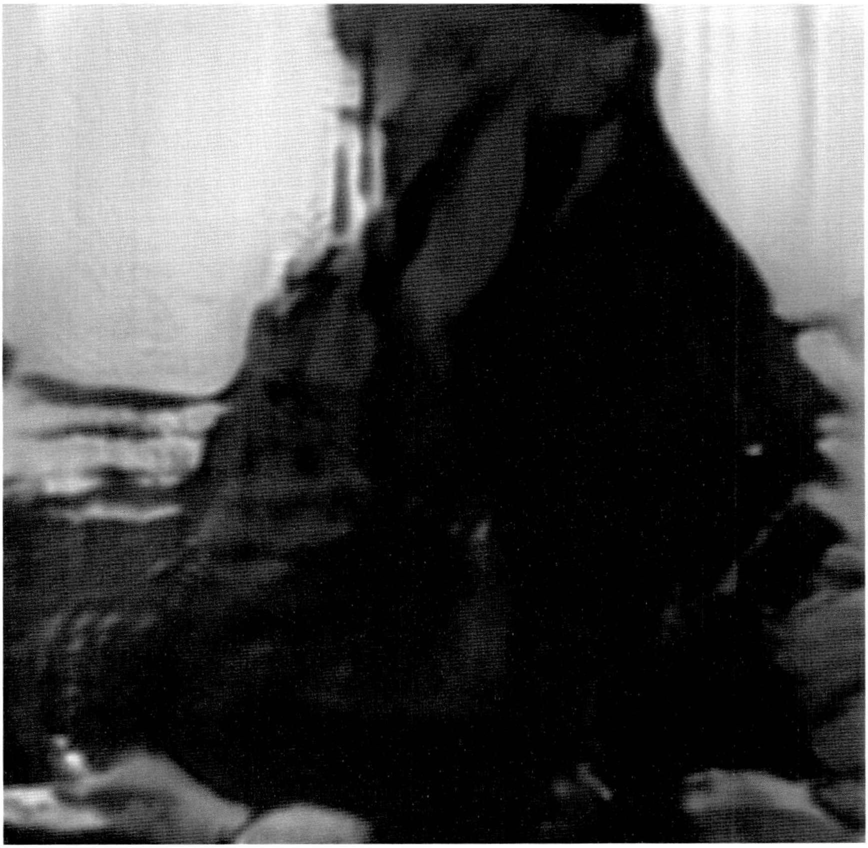

Untitled Film Still 5.18 2019
C-print
16.5 × 16.5 in / 41.9 × 41.9 cm

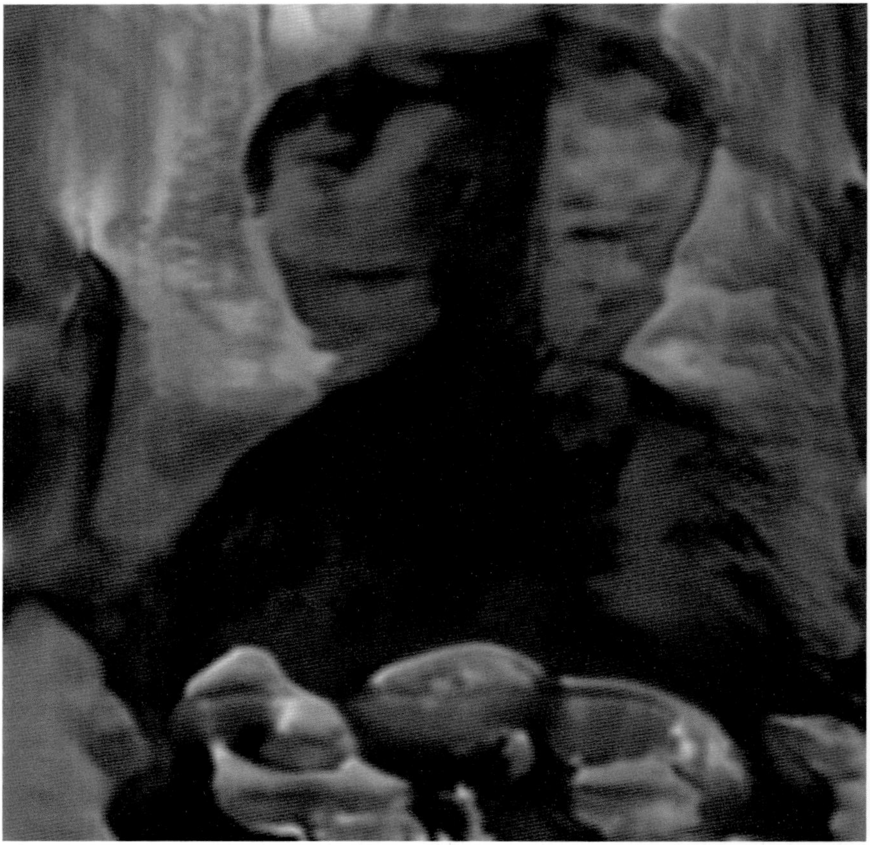

Untitled Film Still 5.19 2019
C-print
16.5 × 16.5 in / 41.9 × 41.9 cm

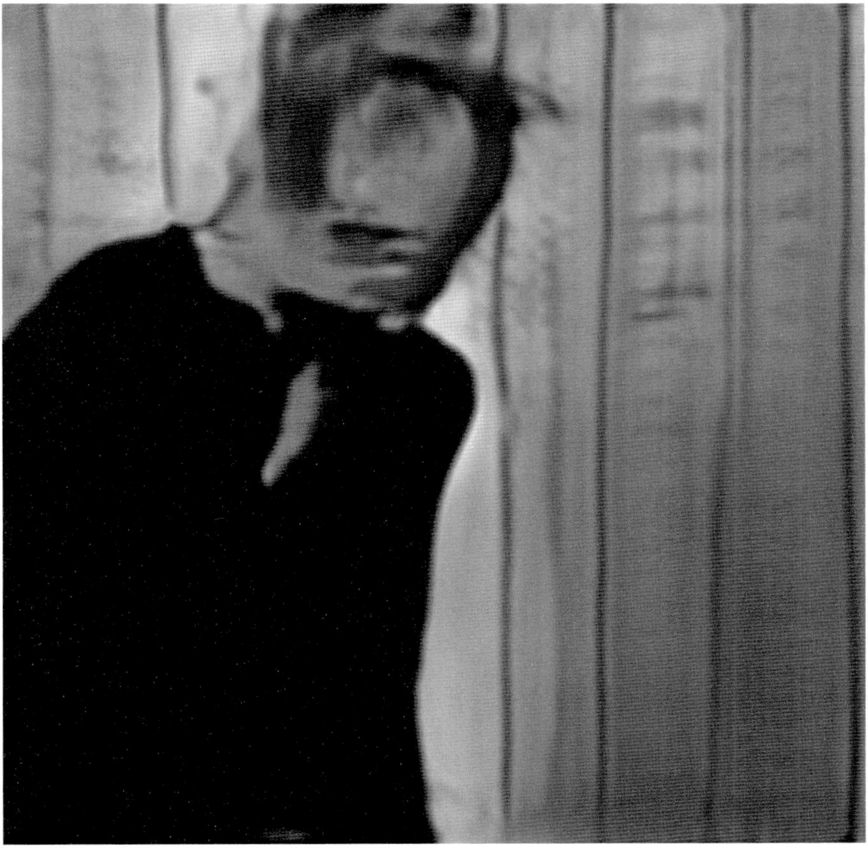

Untitled Film Still 5.20 2019
C-print
16.5 × 16.5 in / 41.9 × 41.9 cm

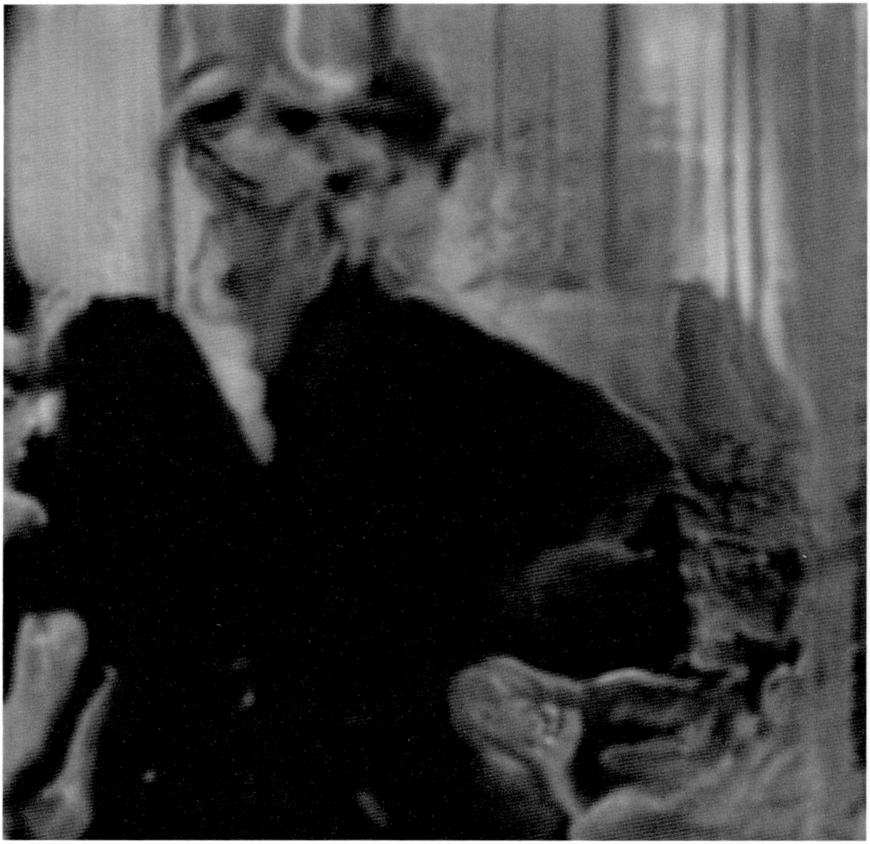

Untitled Film Still 5.21 2019
C-print
16.5 × 16.5 in / 41.9 × 41.9 cm

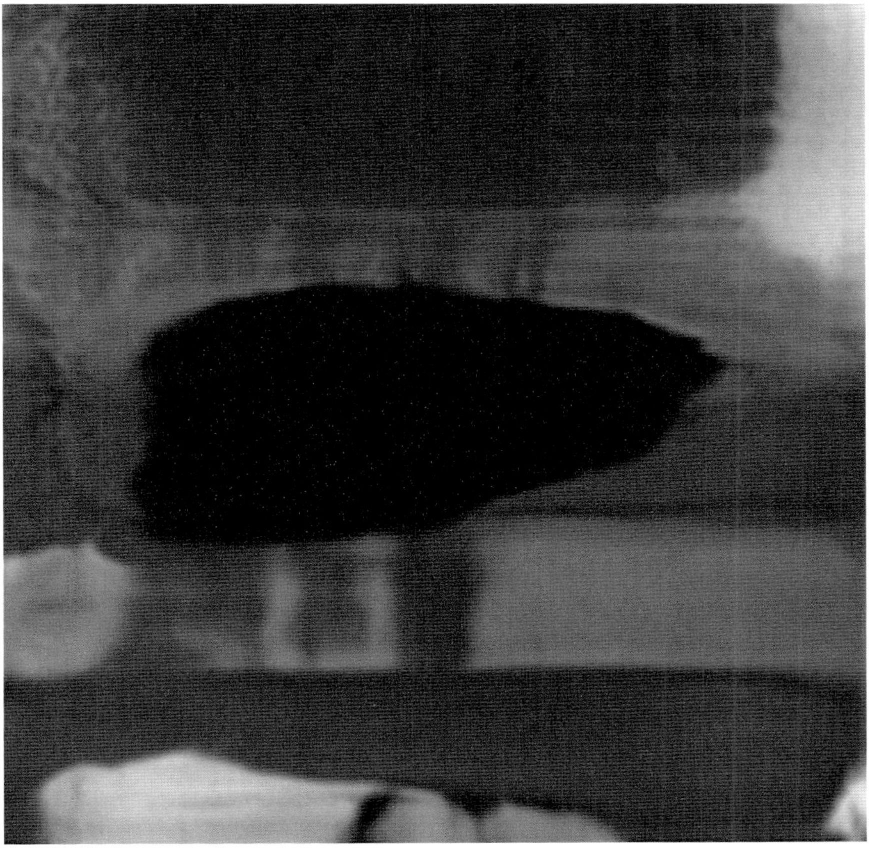

Untitled Film Still 5.22 2019
C-print
16.5 × 16.5 in / 41.9 × 41.9 cm

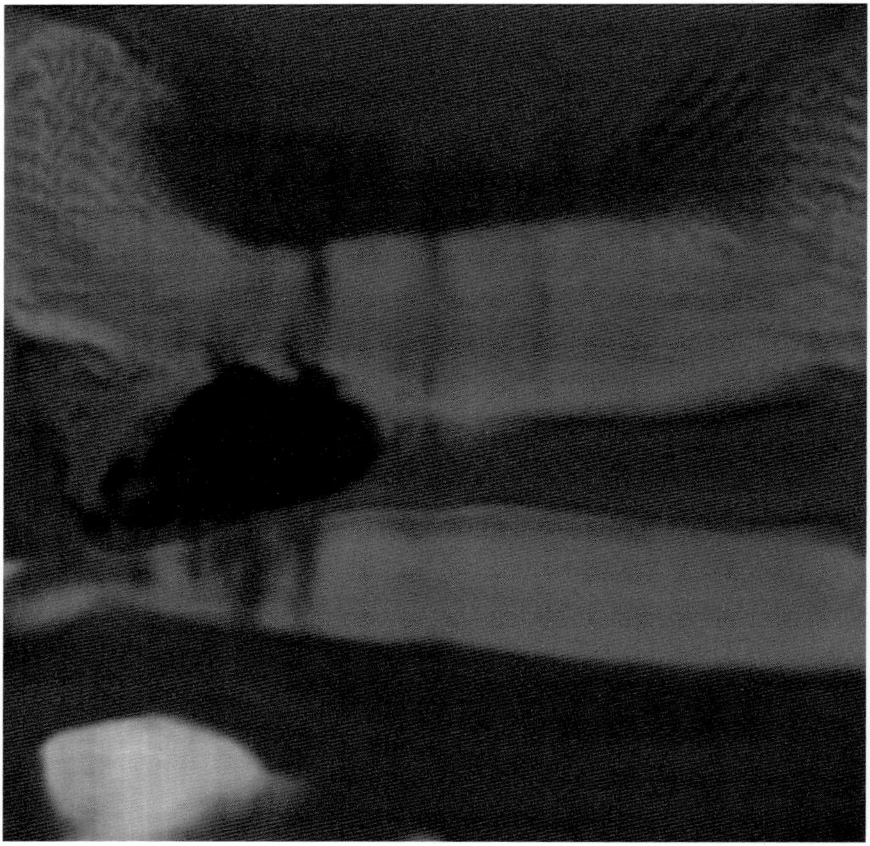

Untitled Film Still 5.23 2019

C-print

16.5 × 16.5 in / 41.9 × 41.9 cm

Biography

Casey Reas is an artist and educator. His software, prints, and installations have been featured in numerous solo and group exhibitions at museums and galleries around the world. His work ranges from small works on paper to urban-scale installations, and he balances solo work in the studio with collaborations with architects and musicians. Reas' work is in a range of private and public collections, including the Centre Georges Pompidou, the San Francisco Museum of Modern Art, and the Whitney Museum of American Art.

Reas is a professor at the University of California, Los Angeles. He holds a master's degree in Media Arts and Sciences from the Massachusetts Institute of Technology, and a bachelor's degree from the College of Design, Architecture, Art, and Planning at the University of Cincinnati. With Ben Fry, Reas initiated Processing in 2001; Processing is an open-source programming language and environment focused on the visual arts.

Reas co-wrote and designed the book *10 PRINT CHR$(205.5+RND(1)); : GOTO 10* (MIT Press, 2013). Reas and Fry published *Processing: A Programming Handbook for Visual Designers and Artists*, (MIT Press, 2007/2014) a comprehensive introduction to programming within the context of visual media. With Chandler McWilliams and LUST, Reas published *Form+Code in Design, Art, and Architecture* (Princeton Architectural Press, 2010), a non-technical introduction to the history, theory, and practice of software in the visual arts. Reas' *Process Compendium 2004–2010* documents six years of his work exploring the phenomena of emergence through software.

Resources for Machine Learning are always changing, we've created an online depository of articles, code and papers at: **anteism.com/casey-reas-resources**

Resources

Working with GANs involves writing code, and it most often involves working with the Python programming language.[7] If this is all new to you, there is a sea of jargon and software systems to navigate. What is Tensorflow? What is Torch? What is Jupyter? Thankfully, there are initiatives to make machine learning approachable for broader audiences such as *Machine Learning for Artists* as *ml5.js* (Friendly Machine Learning for the Web). In my studio, because I wanted to go down a narrow path of working with GANs for making pictures, we developed our own ways of doing things. After trying different options, we settled on working with a dedicated computer running Linux and with a fast GPU.[8] The code we use to make pictures is modified from code posted on Github by researchers.[9] We also write Processing code to develop the thousands of images that are used for training.[10] Without the culture of academic publishing and open-source code, it wouldn't have been possible for me to get started working in this area.

The volume of technical papers released on the topic of machine learning and GANs is daunting. It has taken months to feel comfortable with many of the fundamental ideas and terminology. It is my hope that this short essay can help introduce some of the possibilities and considerations to a wider audience.

[7] Python is a popular programming language used by both beginners and experts. Much of the existing code and tools for machine learning is written with Python.

[8] Specifically, I am running Ubuntu 16.04, CUDA 8.0, and Tensorflow 1.3 with an NVIDIA GeForce GTX 1080 Ti graphics card. A GPU (graphics processing unit) is a specialized circuit that works separately from a computer's main CPU (central processing unit). The GPU is the core of a computer's graphics card, a separate circuit board that accelerates graphics and machine learning code.

[9] Github is an online system for sharing code and collaborating on writing code.

[10] Processing is free, open-source software created for making pictures with code. There's more information at www.processing.org.

Acknowledgements

Hye Min Cho was the technical lead on all of the work documented in this essay. Her expertise and perspicacity were integral to the process. She also read early drafts of this essay and offered valuable feedback. Mike Tyka read the first full draft of the text and provided corrections and ideas that greatly improved the text. Nora Khan's editorial insight sharpened the ideas presented here and I am truly grateful for her expertise.

Editor: Nora N. Khan
Design: Juliette Duhé & Anteism Books
Copy Editing: Anna Prawdzik
Printing & Binding: BookArt, Montréal

ISBN: 978-1-926968-47-6
Second Edition

Artists + Machine Intelligence
ami.withgoogle.com

anteism.com